GROUNDWATER FLOW AND CONTAMINANT TRANSPORT
IN CARBONATE AQUIFERS

Groundwater Flow and Contaminant Transport in Carbonate Aquifers

Edited by

IRA D. SASOWSKY
Department of Geology, University of Akron, Ohio, USA

CAROL M. WICKS
Department of Geological Sciences, University of Missouri, Columbia, Mo., USA

Taylor & Francis
Taylor & Francis Group

LONDON AND NEW YORK

Cover photo: Waterfall in Turkeyscratch Cave, Tennessee. The interplay of vadose and phreatic water in carbonate aquifers leads to complexity not found in granular aquifers. Photo by Ira D. Sasowsky and Ed LaRock.

Published by Taylor & Francis
2 Park Square, Milton Park, Abingdon, Oxon, OX14 4RN
270 Madison Ave, New York NY 10016

Transferred to Digital Printing 2006

ISBN 90 5410 498 8

Publisher's Note
The publisher has gone to great lengths to ensure the quality of this reprint
but points out that some imperfections in the original may be apparent

Printed and bound by CPI Antony Rowe, Eastbourne

Contents

Preface

When soluble rocks such as limestone and dolomite are exposed to the near surface environment, a characteristic landscape called karst develops. This terrane is home to features such as sinking streams, caves, sinkholes, and large springs. From a geomorphic perspective, such settings account for some of the most dramatic landscapes in the world. From a hydrogeologic viewpoint, the carbonate aquifers underlying such regions are some of the most complex and difficult systems to decipher. These aquifers have received increased study in recent years; an interest that has arisen for both pragmatic and intellectual reasons. For the groundwater hydrologist, carbonate aquifers represent a challenging interplay of water flow in conduits, fractures, and pores. To the biologist, they are fragile ecosystems, hosting rare and endangered species. For the geochemist, they may provide rapid transport of contaminants. And to the speleologist, they contain intriguing labyrinths, and deposits of incredible complexity and variety.

The impetus for this book developed in Salt Lake City, Utah, at the 1997 Geological Society of America Annual Meeting. During that meeting a full-day Theme Session on groundwater flow and contaminant transport in carbonate aquifers was held. The session brought together scientists of many different backgrounds with a common interest in karst. Many of the participants have contributed to this volume, along with others who were not initially involved. This has led to a group of papers representing the full spectrum of approaches to studying these systems, with papers ranging from studies that rely on interpretation of field data (dye tracing, chemistry, physical properties), to computer modeling and representation.

Sasowsky's paper begins the volume with an overview of the tools that have been used to understand these aquifers. Studies have historically come from two perspectives, hydrogeological and speleological, but recent work is integrating thoughts from both. Springs are outlet points, and have frequently been used as indicators of carbonate aquifer behavior. The study by Baedke and Krothe relies on dye tracing to springs to interpret the travel times in a karst basin that is contaminated. They used two tracers; Rhodamine WT, which is commonly used by speleologists, and a bromide salt, which is more commonly used by hydrogeologists. In the paper by Wicks and Bohm, a linear systems approach is used to model hydrograph response of a spring to a storm. This led to calculation of a transfer function, which allows prediction of spring response to given storm events. In the following paper, Peterson and others use regression analysis to study the chemical behavior of springs in Arkansas. They found that they could approximate nitrate load based upon stage and specific conductance. Martin and Gordon rely on hydrochemographs to interpret the flow paths and degree of mixing between surface and groundwater in the Floridian aquifer. They found that information about several springs could be gleaned by using simultaneous hydrochemographs of several solutes.

Mace and Hovorka describe three methods that can be used to determine estimates of the porosity and permeability of carbonate aquifers based on detailed studies at the scale of an outcrop. For the Edwards Aquifer of Texas they found that transmissivity can vary by over eight orders of magnitude. Worthington and others examined the relationship between matrix, fracture, and channel storage and flow in Kentucky. They found that matrix accounts for most storage, whereas channels transmit the majority of the water. Velocities of groundwater flow cover nine orders of magnitude in the studied system. Revisiting flow in the Edwards Aquifer, Halihan and others evaluated the contributions of matrix, fracture, and conduit flow. They concluded that fractures control regional scale flow, and that conduits do not usually contribute to well scale permeability. An amazing 15% of wells that were tested showed no measurable drawdown, due to interception of high conductivity zones.

Quinn and Tomasko approached the problem of flow in carbonate aquifers with a modified finite-difference porous medium model. They used a D'arcian approach with drains at the inferred positions of conduits. Calibration at two sites was successful. Zhou's paper on analog and numerical models bridges two very different approaches to modeling karstic basins. He interpreted the results of experiments from three analog models by using a numerical mixing cell model. In the final paper of the volume, Kincaid implemented 3-dimensional data from cave surveys. He presents a method for producing useful maps, including water chemistry data. The eventual integration of such information in to numerical models is likely to lead to more accurate understanding of aquifer behavior.

The papers of this volume serve as a snapshot of carbonate aquifer studies. We thank the authors for their contributions, and look forward to continued progress in this active field.

Ira D. Sasowsky
Carol M. Wicks

Acknowledgments

We are grateful to the numerous professionals who aided in the publication of this book. The following individuals served as reviewers for one or more manuscripts within the volume. Their timeliness and attention to detail have greatly improved the quality of the papers presented herein.

Tom Aley, *Ozark Underground Laboratory*
Ken Bradbury, *Wisconsin Geological and Natural History Survey*
Ralph K. Davis, *University of Arkansas*
Joseph Donovan, *West Virginia University*
James Ewart, *Nittany Geoscience, Inc.*
Malcolm Field, *U.S. Environmental Protection Agency*
John Ganter, *Sandia National Laboratories*
Chris Groves, *Western Kentucky University*
Todd Halihan, *University of Texas - Austin*
Janet Herman, *University of Virginia*
John W. Hess, *Desert Research Institute*
John Hoke, *IT Corporation*
Susan Hovorka, *Bureau of Economic Geology, Texas*
William K. Jones, *Karst Waters Instutiute*
Noel Krothe, *Indiana University*
Robert Mace, *Bureau of Economic Geology, Texas*
June Mirecki, *College of Charleston*
Arthur N. Palmer, *State University of New York - Oneonta*
Eric Peterson, *University of Missouri*
Ward Sanford, *U.S. Geological Survey*
Ira D. Sasowsky, *University of Akron*
George Veni, *George Veni and Associates*
Selena Walko, *University of Akron*
Fred L. Wefer, *Mitre Corporation*
William B. White, *Pennsylvania State University*
Carol M. Wicks, *University of Missouri*
Stephen Worthington, *McMaster University*

Carbonate aquifers:
A review of thoughts and methods

IRA D. SASOWSKY
Department of Geology, University of Akron, Akron, Ohio, USA

ABSTRACT: Water movement in carbonate rocks holds importance both for the purpose of understanding aquifer evolution and from applied standpoints such as water supply and contaminant transport. Conceptual models for these aquifers have historically come from two different perspectives; that of the speleologist and that of the groundwater hydrologist. While conceptual models for the former initially focused on flow in conduits (caves), the latter generally focused on traditional aquifer testing methods. When different tools are used to view the problem, different models are envisioned.

Since the early 1960s, the understanding of water movement in carbonates has become more refined and comprehensive, and the gulf between the two perspectives has narrowed. Present conceptual models are based on data and analyses such as: cave maps, dye traces, pumping tests, remote sensing, spring and stream hydrograph analysis, geochemical studies, surface and borehole geophysics, borehole video, geomorphology, computer modeling, and laboratory experiments. Porosity and flow can occur in at least three physically distinct modes within the rock mass; in the matrix, in fractures, and in conduits. This aquifer framework, along with the types of recharge and discharge, determines the overall behavior of the aquifer. As the scale of the study changes, so may the appropriate conceptual model. Although the identification and characterization of specific flow pathways in a given area is frequently the goal of model development in karst aquifers, it is still difficult or impossible to accomplish in many cases.

1 INTRODUCTION

During the past half century the disciplines of hydrogeology and speleology have each exhibited tremendous development (Domenico & Schwartz, 1998; Meyer et al., 1988; Shaw, 1992; Watson & White, 1985). Both fields have benefited from numerous qualified researchers, a rapidly expanding literature, and a proliferation of field data. There have, however, been major differences in the way in which these two disciplines view the carbonate aquifers that they study. One reason why these differences exist is because each group uses different 'tools' to understand the behavior of the carbonate aquifer. As a generalization, the primary tools of the speleologist were the compass and tape (for mapping caves), and the primary tool of the hydrogeologist was the drill rig. The hydrogeologist pumping an integrated portion of the aquifer might state: 'The carbonate aquifer behaves as a porous medium equivalent'. The speleologist crawling through caverns in the aquifer could counter: 'No, it is dominated by conduit flow – porous medium flow is insignificant'.

Another, and perhaps equally important, reason why these groups have held different views arises from differences in goals for research. Speleologists have traditionally been concerned with the formation of caves and similar questions, whereas hydrogeologists were concerned with finding water. The hydrogeologist evaluated the behavior of the aquifer in the present, but the speleologist was often interested in timescales of thousands or millions of years. These initially disparate views have moved closer with maturation of the science, and the recognition that many of the concerns of these two groups are similar. The study of speleogenesis (cave growth) is equivalent to the study of carbonate aquifer evolution, and the study of karst porosity distribution is of concern to both cave explorers and water suppliers. For these reasons, the terms 'conduit' and 'cave' can be considered equivalent, as can the terms 'carbonate aquifer' and 'karst aquifer'.

The goal of this paper is to provide a background and perspective on the current understanding of carbonate aquifers. This is accomplished through a review of the investigative techniques that have been used, and a historical discussion of some papers which have presented comprehensive conceptual models for these aquifers. The review is designed to be representative, not comprehensive, and many significant references are not included. Papers which are discussed may serve as a point of departure for the reader interested in the many others which unfortunately cannot be discussed in such a brief treatment.

2 INVESTIGATIVE TECHNIQUES

Present conceptual models have been built by the ways in which both hydrogeologist and speleologists have investigated the carbonate aquifer. It is therefore useful to consider what these methods have been, and what they have shown. Within each section below, a brief summary of the technique, along with a limited number of representative papers, is given.

2.1 *Geomorphology*

One of the classic initial works on cave origin was by a geomorphologist (Davis, 1930), and it is fair to say that deductive geomorphology has played a key role in understanding many details of cave development. Early texts about karst had a distinctly geomorphic focus (Jennings, 1971; Sweeting, 1973). Morphology has been useful on three scales: specific features within the cave, the overall form of the cave (addressed in next section), and the cave in relation to its environment. Within the cave, the shape of the conduit aids in determining whether the cave formed in the phreatic or vadose zone. Using dissolutional scallops on the walls, along with passage cross-sectional area, computations of paleodischarge through the portion of the aquifer may be made (Curl, 1974). By evaluating the cave in the context of its geologic setting, an understanding of groundwater flow patterns can be deduced (Palmer, 1984, 1991).

R.L. Curl has been concerned with the distribution of caves and their shapes. He found that caves are fractal objects, and that they follow a distribution having a very large number of short caves, and a very small number of long caves. In addition, the long caves tend to have a greater number of entrances (Curl, 1958, 1966, 1986).

2.2 *Cave maps*

Mapping and illustration are critical for the geosciences because it is difficult to understand something that cannot be visualized. Caves share the same difficulty found with other geologic features (faults, anticlines, etc.) in that they must be mapped to be understood. The proliferation of cave maps, perhaps more than anything else, has led to progress in understanding the distribution and modes of karst porosity. The number of maps in all parts of the world increases rapidly each year, mainly due to the efforts of amateur speleologists using hand-held compasses and fiberglass tapes. Most maps are not published in scientific journals, but are found in local caving club newsletters, as regional compilations (Douglas, 1964; Medville & Medville, 1995), or as global studies (Courbon et al., 1989). The wide variety of patterns that caves follow has provoked the investigation of factors controlling their development. For example, by using published data on thousands of caves, and personal observations in over 500, Palmer (1991) classified cave patterns in relation to rock porosity and recharge type (Fig. 1).

2.3 *Groundwater tracing*

Groundwater tracing, commonly called 'dye tracing', is the process of placing a soluble or buoyant material in an aquifer (frequently in the recharge zone), and seeing where it

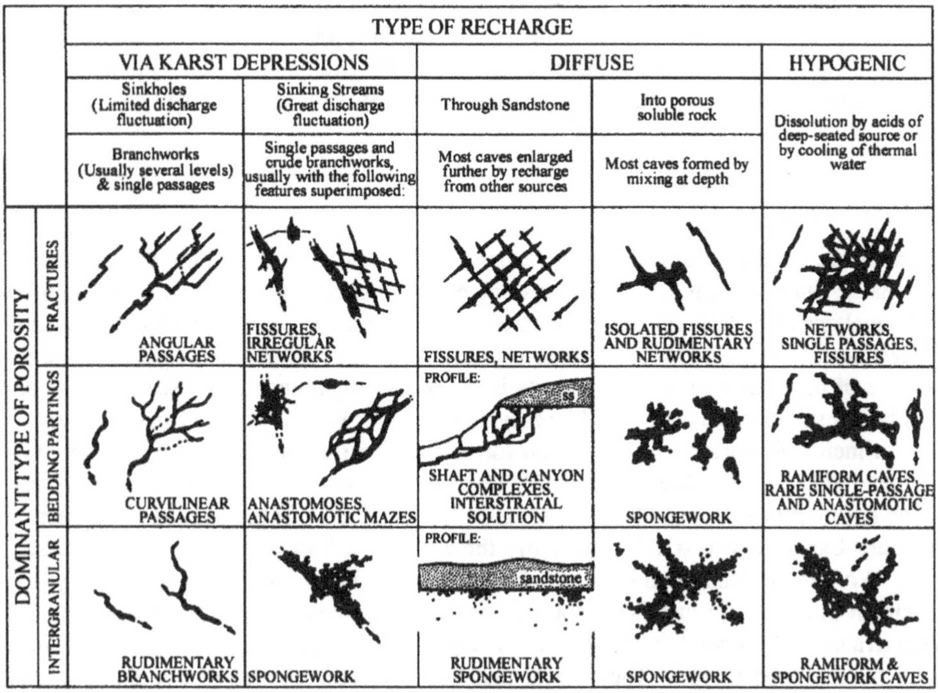

Figure 1. Summary of cave patterns and relationships to recharge and porosity, from Palmer, 1991.

comes out. This has been conducted (sometimes accidentally) since historic times using materials as diverse as chaff, whiskey, bombs, and ducks. The most commonly used materials are dyes (Davis et al., 1980; Gaspar, 1987; Jones, 1984b; Smart & Laidlaw, 1977).

This technique provides answers to the basic questions about what parts of the aquifer are connected to others, which direction the water is moving, and how fast it is moving. Initial use of tracing focused on karst conduits, where the inputs to the aquifer (sinkholes, swallets) were very distinct, as were the outputs (springs). Tracers were used to determine specific connections because the routes were not humanly traversable (either due to size or flooding). In addition to answering the above basic questions, tracing studies have allowed the delineation of karst groundwater basins (Gospodariè & Habiè, 1976; Jones, 1973; Quinlan & Ewers, 1989). In carbonate aquifers it is quite common for the groundwater basin to not coincide with the surface water basin. There is considerable complexity and difficulty with interpretation. Basin divides may shift with changing groundwater stage, and basins may actually overlap each other in plan view.

Increased concern with groundwater contaminant migration has made dye tracing an important tool for the identification of specific flow paths, both in carbonate and non-carbonate aquifers. This includes identification of potential migration routes, as in assessing suitability of a site for a particular use such as landfilling (Aley, 1988), or the confirmation of sources of known contamination (Garman & Fischer, 1988). Tracing between boreholes, and also at multiple zones within wells, has become widely used in fractured insoluble media. The implementation of quantitative tests, with semi-continuous monitoring of dye recovery, allows for calculation of additional information about the aquifer being studied (Jones, 1984a).

Regulatory requirements for groundwater evaluation initially discouraged the use of springs and dye tracing for environmental projects in carbonate areas. This has been changing, with the increased recognition that special protocols are required for useful investigation of these aquifers (Eckenfelder Inc., 1996; Mull et al., 1988; Quinlan, 1989). Tracing to a well is difficult, but may be successful, if the well is pumped at such a rate as to induce flow from a large area.

2.4 *Hydrograph analysis*

Springs are natural outlets for groundwater in karst aquifers, and temporal variations in discharge can provide insight in to the storage and transport characteristics of the aquifer. The analysis of hydrographs has been frequently used in conjunction with either quantitative dye tracing or variation in some chemical parameter (hydrochemographs). The response of the system to recharge events is of highest importance to the speleologists actively investigating the aquifer – they may be trapped or killed by rapidly rising waters.

An incredible variety of hydrograph forms are seen from karst springs (Ford & Williams, 1989, p. 194). These reflect the diverse ways in which carbonate aquifers can evolve. Smart (1983) examined spring response to flooding in the Cathedral Aquifer of western Canada and was able to identify three different flow systems in this alpine setting. Work in the Mammoth Cave aquifer of Kentucky has been ongoing for many years. Meiman et al. (1988) installed dataloggers recording stage, velocity, conductivity, and temperature at four conduit sites within the aquifer. Readings were made every two minutes for extended periods. They found that flood pulses consisted of two components; displaced storage water, and fresh recharge. These were differentiated on both chemical and physical characteristics. Building upon work by Neuman & de Marsily (1976), Dreiss (1982) developed the concept of applying linear kernel functions to large springs, and found that this could predict the response of a spring in some circumstances. This was extended to include separation of flow components and comparison of regional scale transport in Missouri (Dreiss, 1989a,b). Studying the Berghan Spring of southern Iran, Raeisi & Karami (1997) separated baseflow from the total spring output using chemographs, and concluded that flow was dominantly diffuse in that aquifer.

2.5 *Pumping tests*

Perhaps more than any other investigative technique, the use of pumping tests has been a point of debate. At issue is the validity of using analytical methods designed for porous medium (D'arcy) materials in carbonate aquifers, which in most cases have components of flow which are non-D'arcian. For determining aquifer parameters on a broad scale, such testing is appropriate many times. However, for determining contaminant transport pathways and velocities it rarely is. Good insight into aquifer behavior has been gained by conducting such tests. In the folded carbonate rocks of the Appalachians (Parizek et al., 1971, p.93) there is a strong correlation between well productivity and proximity to fracture traces (fracture zones). This is caused by high conductivity both in original and enlarged (karstified) fractures. Specific methods for extracting aquifer parameters have been proposed for carbonate aquifers (Mace, 1997; Thrailkill, 1988). In a 'slug' test (Michalski & Torlucci, 1988) using storm water input to a sinkhole in the North Coast of Puerto Rico, it was believed that the aquifer demonstrated near D'arcian response, though this was debated (Ewers et al., 1989).

2.6 *Geochemistry*

The process of carbonate dissolution exerts the most apparent geochemical effect on the behavior of carbonate aquifers by controlling groundwater flow. The type and distribution of porosity in the aquifer are dictated by the geochemical history of the rock. For the most part, karst porosity develops when meteoric waters become enriched with CO_2, forming carbonic acid (H_2CO_3), which is aggressive to calcite. Dissolution occurs through a series of reactions that have been the subject of many field, laboratory, and theoretical investigations.

The recognition that time variation in water chemistry was different between karst springs, even in a given area, led to questions about flow behavior within carbonate aquifers. Shuster & White (1971) considered that springs showing a high coefficient of variation for hardness (as well as temperature and calcite saturation index) drained aquifers dominated by conduit flow, whereas those with low variation drained aquifers where flow was diffuse (through small fractures). Others have concluded (Worthington et al., 1992) that the bulk of variation is explained by recharge to the aquifer, rather than flow style within the rock mass.

A particular problem of cave formation was posed by speleologists in the 1960s. It was recognized that most CO_2, and hence acidity, was imparted to groundwater in the soil zone. With this being the case, how could caves form at substantial depth if their dissolving power would be expended near the surface, during initial contact with limestone?

Mixing corrosion was offered as a possible mechanism (Bögli, 1964). It is caused by the non-linear relationship between calcite solubility and CO_2, concentration (Plummer & Wigley, 1976), that allows for two waters saturated with respect to calcite to mix and become aggressive. Another possible mechanism is the deep generation of CO_2 by microbial oxidation of carbon (Wood, 1985; Wood & Petraitis, 1984). Using field and laboratory data, it has been found that slightly aggressive meteoric water may persist to great depth in limestone because the reaction kinetics become quite sluggish at conditions near equilibrium (White, 1977b). Combining this work with flow modeling, it has been shown that this 'kinetic triggering' mechanism can account for most cave formation (Palmer, 1991).

The hydrochemistry of coastal limestone regions has received special interest. Work by Back et al. (1979) established a relationship between coastal landform evolution and fresh groundwater discharge, where increased dissolution via mixing corrosion was taking place. Mylroie et al. (1995) developed a model of Flank Margin Cave Development to explain the occurrence and morphology of caves found on carbonate islands. These caves form in discharge zones where fresh and salt water mix, and are found at various elevations due to changes in sea level.

2.7 *Remote sensing and surface geophysics*

Remote sensing includes such airborne surveys as visible and infrared photography, spectral scanning, and thermal scanning. These have shown good success in conducting broad surveys of areas to identify karst features such as springs and sinking streams (e.g. Bogle & Loy, 1995), but have not contributed directly to the development of models for carbonate aquifers. Surface geophysics have been put to great use in applied studies concerned for locating buried voids and sinkholes (Benson et al., 1995; Lambert, 1997; Luke & Chase, 1997). In areas underlain by carbonate rock, electromagnetic, gravity, and ground penetrating radar surveys are frequently conducted by consulting firms to meet specific client needs. Although such surveys have the potential to provide important information on the recharge zones of carbonate aquifers, there has been little attempt to integrate such information into the overall function of the aquifer.

2.8 *Borehole geophysics*

Being able to 'see' the aquifer has been an advantage held by speleologists, but it provides a biased sample – showing only the conduit porosity large enough for humans to enter. Borehole geophysical methods allow the observation of other parts of the aquifer, which may or may not include conduits. Use of these techniques has been limited as a result of the expense and difficulty of using these tools, but their use is becoming more commonplace. Classical devices such as caliper, gamma ray, density, and porosity, which were designed for use in petroleum development, are all useful in groundwater investigations, revealing porosity and permeability trends (Benson & Yuhr, 1993).

Perhaps the most exciting tool has been the borehole video camera. Initially these were extremely expensive devices giving a downward-looking black-and-white image requiring a large borehole. Newer models are inexpensive, provide full color images in 5 cm (2 inch) wells, and are equipped with side-looking capability to allow views directly in to fractures and conduits. Inclusion of a compass on the unit allows accurate surveys of fracture and conduit orientation and aperture.

2.9 *Computer modeling*

Modeling of carbonate aquifers has focused on three aspects: numerical models of present day groundwater flow, geochemical modeling of aqueous reactions taking place within the aquifer, and growth models which combine aspects of flow and geochemical models.

Numerical flow models have seen the most use because they are frequently required for solving applied problems. Models such as MODFLOW (McDonald & Harbaugh, 1988) are designed for application in porous media (D'arcian conditions) but have been adapted with varying success for use in some karst settings. Such application requires the use of drains and high conductivity layers within the bulk of the aquifer. One of the earliest models implemented specifically in a karst setting was a pipe-flow (non-D'arcian) model for the Sinkhole Plain Aquifer near Mammoth Cave, Kentucky (Thrailkill, 1972). Current efforts in modeling for karst aquifers are an outgrowth of the great interest in modeling groundwater flow in fractured (non-soluble) media (e.g. Billaux & Gentier, 1990; Chilès et al., 1992; Neuman, 1997; Pyrak-Nolte et al., 1990). One recent compilation listed 28 commercially or academically available computer programs for such modeling (Diodato, 1994). Approaches taken include treating the aquifers as a porous medium (with or without extreme specification of boundary conditions), as explicit discrete fractures, as a dual continuum, or as discrete fracture networks. The common difficulty with all of these applications is the limited amount of information about fracture and conduit position, conductivity, and interconnection. Such models are likely to see refinement and increased complexity, as more powerful computer hardware and software becomes available each year.

Geochemical models such as WATEQ4F (Ball & Nordstrom, 1991) have been used to remove the tedium of calculating rock/mineral interactions. Models such as NETPATH (Plummer et al., 1991) define reaction models consistent with given data whereas models such as PHREEQE (Parkhurst et al., 1980) are useful for applying an assumed reaction to some given initial conditions. All give insight into the geochemical processes acting upon the aquifer through time, or through space.

From the speleological viewpoint, there is great interest in joining flow models with reactive geochemical models to simulate cave growth (Dreybrodt, 1992; Groves & Howard, 1994; Palmer, 1991). These have been essential to a genetic scheme for carbonate aquifers, and help in understanding the observed porosity distribution.

2.10 *Laboratory experiments*

With the exception of geochemical studies, laboratory work concerned with carbonate aquifers has been minimal. It has, however, been important for constraining models of conduit development. Numerous papers have focused on dissolution rates of minerals under varying conditions (Plummer et al., 1979; Plummer & Wigley, 1976). Rauch and White examined dissolution rates of various carbonate units and compared them to lengths of known cave (Rauch & White, 1970, 1977). Actual physical models for the growth of conduits have been assembled using rapidly soluble materials (Ewers, 1982).

3 HISTORICAL DEVELOPMENT OF CONCEPTUAL MODELS

Conceptual models are important because they control the way in which we investigate and design experiments. For example, identification of an appropriate conceptual model is the first step in development of a mathematical ground water flow model (Mercer & Faust, 1980). Oftentimes a conceptual model goes unstated, being reflected only as a way of thinking, or in a method for solving a problem. For the case of carbonate aquifers, numerous papers have helped to build portions of a general conceptual model. In addition, four papers have presented comprehensive conceptual models for carbonate aquifers. Three of these conceptualized present-day aquifer flow behavior, while the other was genetic, conceptualizing conduit growth through time.

3.1 *Specific concepts*

In addition to those papers mentioned in the preceding section on investigative techniques, other papers have been particularly significant for developing ideas in support of an overall conceptual model. Some of these include: Stringfield (1936) for regional evaluation of a major karst aquifer; Williams (1983) for flow within the uppermost vadose (recharge) zone; Parizek et al. (1971) for evaluating the effects of fracturing and karstification on well yields; Ford & Ewers (1978) for relating conduit development to fracture characteristics; and Teutsch & Sauter (1992) for examination of scaling effects on aquifer behavior. The reader is also referred to several excellent textbooks and monographs (e.g., Dreybrodt, 1988; Ford & Williams, 1989; White, 1988) that provide overviews of karst geoscience.

3.2 *Comprehensive conceptual models*

The first of the comprehensive papers (White, 1969) was concerned with flow behavior, and visualized that there existed more than one component of flow within the carbonate aquifer. These included a diffuse (laminar) flow occurring within un-enlarged fractures in the rock, and conduit (turbulent) flow present in enlarged fractures. This concept, and the terms 'diffuse' and 'conduit', led to many examinations of what factors actually controlled the behavior and character of groundwater flow in carbonates.

Building upon this model, White (1977a) developed a scheme tying the overall setting of the rock mass to structural and geologic controls. The question of scale (overall aquifer size) was also addressed, with the recognition that free-flowing conduit systems are likely to develop in catchments on the order of 10^0 to 10^2 km^2, and that diffuse flow and artesian aquifers were more likely in larger areas.

Smart and Hobbs (1986) viewed the aquifer as a system characterized by the three variables of recharge, flow and, storage (Fig. 2). Within the aquifer, recharge can be either concentrated (as through sinkholes and sinking streams) or dispersed (as through a multitude of areally distributed fractures). Flow can be either conduit (turbulent, through solutionally enlarged openings) or diffuse (laminar, through small pores and fractures). Storage can be of a variety of types, with general end members being unsaturated (vadose) and saturated (phreatic). In essence, this system incorporates the classical hydrogeologic concepts of permeability, storativity, and boundary conditions (recharge/discharge zones).

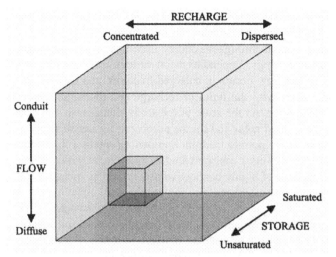

Figure 2. Conceptual model of the carbonate aquifer as a continuum of recharge, flow, and storage types (after Smart & Hobbs, 1986). Smaller cube represents a range of behavior for one hypothetical aquifer.

Palmer (1991) synthesized earlier works and combined them with his own extensive field observations and digital geochemical modeling. From this he developed a comprehensive genetic model of cave origin. A time period of 10,000 to 100,000 years is needed under most conditions to develop large cavernous porosity in an aquifer. By characterizing geologic setting, along with hydrologic boundary conditions, one may predict expected cave formation and aquifer behavior.

4 DISCUSSION AND SUMMARY

Given the multitude of techniques discussed earlier, it is understandable that investigators using different tools have developed differing views of the behavior of carbonate aquifers. It is satisfying that the fusion of information from these techniques is resulting in general conceptual models that accurately reflect aquifer behavior in a range of settings, and at a range of scales.

Much of the information generated has come from the need to address applied problems in these aquifers. This has cast together hydrogeologists and speleologists who previously might not have had common interests. The perspective of engineers who are called upon to investigate and to implement solutions for environmental problems has also been called to bear. Many successful collaborations have been reported in a series of conferences sponsored by the National Ground Water Association, The Florida Sinkhole Research Institute, and P.E. LaMoreaux and Associates (Beck, 1984, 1989, 1993, 1995; Beck & Stephenson, 1997; Beck & Wilson, 1987; Dilamarter & Csallany, 1977; National Ground Water Association, 1991; National Water Well Association, 1986, 1988). As additional data become published, future conceptual models are certain to become more refined.

It has been accepted by most scientists that the presence of carbonate rocks implies karstification, but that the character of karst development in different settings is highly variable. Porosity, and hence flow, can occur in 3 interlinked modes. Matrix flow moves through intergranular (primary) pores, and is characterized by diffuse or D'arcian flow. Fracture flow moves in a laminar manner through secondary openings in the rock, such as joints, faults, and bedding plane partings. Depending upon various conditions it may also be referred to as diffuse. Conduit flow occurs in enlarged fractures, and may be turbulent. This aquifer framework, along with the types of recharge and discharge, determines the overall behavior of the aquifer. As the scale of the study changes, so may the appropriate conceptual model. Over broad areas and for the purposes of water supply, for example, it may be appropriate to apply porous medium approximations to a limestone aquifer. For smaller areas, and for contaminant transport studies, however, it would rarely be appropriate to do so; a small number of highly conductive pathways may transport the vast majority of flow.

Further advances in understanding are likely to come from the use of computers for processing, modeling, and visualizing the vast amounts of data that have been and will be collected. In particular, the use of visualization software is likely to lead to the recognition of relationships between groundwater flow and geologic properties. The entire science of hydrogeology operates under the handicap of not being able to directly 'see' what is occurring in the aquifer. The ongoing proliferation of computer software and increasingly powerful hardware will allow some measure of improvement in this regard.

The identification and characterization of specific flow pathways at a given site is probably the most sought after (and difficult) outcome in evaluating a karst aquifer. Such characterization might allow, for example, exact placement of extraction wells to remove contamination from the aquifer. Future advances in remote sensing and modeling will bring the science closer to achieving such precise evaluation capability.

ACKNOWLEDGMENTS

This work is a synthesis of ideas that others have had, so it is most appropriate to acknowledge the scientists (both cited and uncited) who have studied carbonate aquifers. These include colleagues in the Geological Society of America, the National Ground Water Association, and the National Speleological Society. Janet Herman and one anonymous reviewer provided very useful reviews of an initial version of this paper. Thanks to Kathy Sasowsky and Selena Walko for other helpful editorial comments.

REFERENCES

Aley, T. 1988. Complex radial flow of ground water in flat-lying, residuum-mantled limestone in the Arkansas Ozarks, in *Proceedings of the Second Conference on Environmental Problems in Karst Terranes and Their Solutions Conference, Nashville*, National Water Well Association, p. 159-170.

Back, W., Hanshaw, B.B., Pyle, T., Plummer, L.N., & Weidie, A.E., 1979. Geochemical significance of groundwater discharge and carbonate solution to the formation of Caleta Xel Ha, Quintana Roo, Mexico: Water Resources Research, v. 15, p. 1521-1535.

Ball, J.W. & Nordstrom, D.K. 1991. User's manual for WATEQ4F, with revised thermodynamic databse and test cases for calculating speciation of major, trace, and redox elements in natural waters: United States Geological Survey, Open-File Report 91-183, 191 p.

Beck, B.F. 1984. Sinkholes: Their geology, engineering, and environmental impact. *Proceedings of the First Multidisciplinary Conference on Sinkholes*. Rotterdam: A.A.Balkema, 429 p.

Beck, B.F. 1989. Environmental and engineering impacts of sinkholes and karst. *Proceedings of the Third Multidisciplinary Conference on Sinkholes and the Engineering and Environmental Impacts of Karst*. Rotterdam: A.A.Balkema, 384 p.

Beck, B.F. 1993. Applied karst geology. *Proceedings of the Fourth Multidisciplinary Conference on Sinkholes and the Engineering and Environmental Impacts of Karst*. Rotterdam: A.A.Balkema, 295 p.

Beck, B.F. 1995. Karst Geohazards. *Proceedings of the Fifth Multidisciplinary Conference on Sinkholes and the Engineering and Environmental Impacts of Karst. Gatlinburg, Tennessee*. Rotterdam: A.A.Balkema, 581 p.

Beck, B.F. & Stephenson, J.B. 1997. The engineering geology and hydrogeology of karst terranes. *Proceedings of the Sixth Multidisciplinary Conference on Sinkholes and the Engineering and Environmental Impacts of Karst*. Rotterdam: A.A.Balkema, 516 p.

Beck, B.F. & Wilson, W.L. 1987. Karst Hydrogeology: Engineering and environmental applications. *Proceedings of the Second Multidisciplinary Conference on Sinkholes and the Environmental Impacts of Karst*. Rotterdam: A.A.Balkema, 467 p.

Benson, R.C. & Yuhr, L. 1993. Spatial sampling considerations and their applications to characterizing fractured rock and karst systems, in Beck, B.F. (ed.) *Applied Karst Geology*. Rotterdam: A.A.Balkema, p. 99-113.

Benson, R.C., Yuhr, L. & Berkovitz, B.C. 1995. Subsurface investigation of possible karst conditions at the Jewfish Creek bridge replacement, Key Largo, Florida, in Beck, B.F. (ed.) Karst Geohazards. *Proceedings of the Fifth Multidisciplinary Conference on Sinkholes and the Engineering and Environmental Impacts of Karst*. Rotterdam: A.A. Balkema, p. 409-414.

Billaux, D. & Gentier, S. 1990. Numerical and laboratory studies of flow in a fracture, in Barton, N. & Stephansson, O. (eds) *Rock Joints: Proceedings of the International Symposium on Rock Joints*. Rotterdam: A.A. Balkema, p. 369-373.

Bögle, F.R. & Loy, K. 1995. The application of thermal infrared thermography in the identifcation of submerged springs in Chicakamauga Reservoir, Hamilton County, Tennessee, in Beck, B.F. (ed.) Karst Geohazards. *Proceedings of the Fifth Multidisciplinary Conference on Sinkholes and the Engineering and Environmental Impacts of Karst*. Rotterdam: A.A. Balkema, p. 415-424.

Bögli, A. 1964. Mischkungskorrosion – ein Beitrang zum Verkarstungproblem. *Erdkunde*, 18: 83-92.

Chilès, J.P., Guérin, F. & Billaux, D. 1992. 3D stochastic simulation of fracture network and flow at Stripa conditioned on observed fractures and calibrated on measured flow rates, in Tillerson, J.R. & Wawersik, W.R. (eds) *Rock Mechanics: Proceedings of the 33rd US symposium*. Rotterdam: A.A. Balkema, p. 533-542.

Courbon, P., Chabert, C., Bosted, P. & Lindsley, K. 1989. *Atlas of the great caves of the world*. St. Louis: Cave Books, 369 p.

Curl, R.L. 1958. A statistical theory of cave entrance evolution. *National Speleological Society Bulletin*, 20: 9-22.

Curl, R.L. 1966. Caves as a measure of karst. *Journal of Geology*, 74: 798-830.

Curl, R.L. 1974. Deducing flow velocity in cave conduits from scallops. *National Speleological Society Bulletin*, 36(2): 1-5.

Curl, R.L. 1986. Fractal dimensions and geometries of caves: Mathematical Geology, 18: 765-783.

Davis, S.N., Thompson, G.M., Bentley, H.W. & Stiles, G. 1980. Ground-Water tracers – a short review. *Ground Water*, 18(1): 14-23.

Davis, W.M. 1930. Origin of limestone caverns. *Geological Society of America Bulletin*, 41: 475-628.

Dilamarter, R.R. & Csallany, S.C. 1977. Hydrologic problems in karst regions. Bowling Green: Western Kentucky University, 481 p.

Diodato, D.M. 1994. A brief compendium of fracture flow models: Chicago, Argonne National Laboratory, Energy Systems Division, 60 p. Also available at http://www.ems.psu.edu/Hydrogeologist/docs/cffm/cffm.pdf

Domenico, P.A. & Schwartz, F.W. 1998. *Physical and chemical hydrogeology*. New York: John Wiley & Sons, 506 p.

Douglas, 1964. *Caves of Virginia*. Falls Church, Virginia: Virginia Cave Survey, 761 p.

Dreiss, S.J. 1982. Linear kernels for karst aquifers. *Water Resources Research*, 18(4): 865-876.

Dreiss, S.J. 1989a. Regional scale transport in a karst aquifer 1. Component separation of spring flow hydrographs. *Water Resources Research*, 25(1): 117-125.

Dreiss, S.J. 1989b. Regional scale transport in a karst aquifer 2. Linear systems and time moment analysis. *Water Resources Research*, 25(1): 126-134.

Dreybrodt, W. 1988. Processes in karst systems. *Physics, chemistry, and geology, Springer series in physical environment – 4*. Berlin: Springer-Verlag, 288 p.

Dreybrodt, W. 1992. Dynamics of karstification: A model applied to hydraulic structures in karst terranes. *Applied Hydrogeology*, 1(3): 20-32.

Eckenfelder Inc. 1996. Guidelines for wellhead and springhead protection area delineation in carbonate rocks. United States Environmental Protection Agency, Region 4, EPA 904-B-97-003, numbers), v. p.

Ewers, R.O. 1982. Cavern development in the dimensions of length and breadth (Ph.D. thesis). McMaster University, 398 p.

Ewers, R.O., Keagy, D., Quinlan, J.F. & Field, M. 1989. *Discussion of "'Testing' a limestone aquifer using water-table response to stormwater discharged into sinkholes"*, by Andrew Michalski & Joseph Torlucci, Jr. *Ground Water*, 27(5): 715-716.

Ford, D.C. & Ewers, R.O. 1978. The development of limestone caverns in the dimensions of length and depth: Canadian Journal of Earth Sciences, 15: 1783-1798.

Ford, D.C. & Williams, P.W. 1989. Karst geomorphology and hydrology. London: Unwin Hyman, 601 p.

Garman, P.M. & Fischer, F.T. 1988. A landfill/ground-water contamination case study. *Proceedings of the Second Conference on Environmental Problems in Karst Terranes and Their Solutions Conference*. Nashville: National Water Well Association, p. 143-158.

Gaspar, E. 1987. *Modern trends in tracer hydrology*. Boca Raton: CRC Press, 139 p.

Gospodariè, R. & Habiè, P. 1976. Underground water tracing. *Investigations in Slovenia 1972-1975*. Postona: Institute for Karst Research SAZU, 309 p.

Groves, C.G. & Howard, A.D. 1994. Early development of karst systems 1. Preferential flow path enlargement under laminar flow. *Water Resources Research*, 30(10): 2837-2846.

Jennings, J.N. 1971. Karst, An introduction to systematic geomorphology. Cambridge: Massachusetts, M.I.T. Press, 252 p.

Jones, W.K. 1973. Hydrology of limestone karst in Greenbrier County, West Virginia. *West Virginia Geological Survey, Bulletin 36*, 49 p.

Jones, W.K. 1984a. Analysis and interpretation of data from tracer tests in karst areas. *NSS Bulletin*, 46: 41-47.

Jones, W.K. 1984b. Dye tracer tests in karst areas. *National Speleological Society Bulletin*, 46: 3-9.

Lambert, D.W. 1997. Dipole-dipole D.C. resistivity surveying for exploration of karst features, in Beck, B.F. & Stephenson, J.B. (eds) The engineering geology and hydrogeology of karst terranes. *Proceedings of the Sixth Multidisciplinary Conference on Sinkholes and the Engineering and Environmental Impacts of Karst*. Rotterdam: A.A. Balkema, p. 413-418.

Luke, B.A. & Chase, D.S. 1997. Detecting caves using seismic waves: A feasibility study, in Beck, B.F. & Stephenson, J.B. (eds) The engineering geology and hydrogeology of karst terranes. *Proceedings of the Sixth Multidisciplinary Conference on Sinkholes and the Engineering and Environmental Impacts of Karst*. Rotterdam: A.A. Balkema, p. 419-424.

Mace, R.E. 1997. Determination of transmissivity from specific capacity tests in a karst aquifer: Ground Water, 35(5): 738-742.

McDonald, M.G. & Harbaugh, A.W. 1988. A modular three-dimensional finite-difference ground-water flow model. *United States Geological Survey, Techniques of Water Resources Investigations*. Book 6, Chapter A1, unnumbered.

Medville, D. & Medville, H. 1995. Caves and karst of Randolph County (West Virginia).: *West Virginia Speleological Survey, Bulletin 13*, 250 p.

Meiman, J., Ewers, R.O. & Quinlan, J.F. 1988. Investigation of flood pulse movement through a maturely karstified aquifer at Mammoth Cave National Park: A new approach, in National Water Well Association (ed.) *Proceedings of the Second Conference on Environmental Problems in Karst Terranes and Their Solutions*. Nashville: National Water Well Association, p. 227-262.

Mercer, J.W. & Faust, C.R. 1980. Ground-water modeling: An overview. *Ground Water*, 18(2): 108-115.

Meyer, G., Davis, G. & LaMoreaux, P.E. 1988. Historical perspective, in Back, W., Rosenshein, J.S. & Seaber, P.R. (eds) *Hydrogeology: The geology of North America, O-2: Boulder, Geological Society of America*, p. 1-8.

Michalski, A. & Torlucci, J. Jr. 1988. 'Testing' a limestone aquifer using water-table response to stormwater discharged into sinkholes. *Ground Water*, 26(6): 751-760.

Mull, D.S., Lieberman, T.D., Smoot, J.L. & Woosley, L.H. Jr. 1988. *Application of dye-tracing techniques for determining solute transport characteristics of ground water in karst terranes*. United States Environmental Protection Agency, Region 4, EPA 904-/6-88-001, 103 p.

Mylroie, J.E., Carew, J.L. & Vacher, H.L. 1995. Karst development in the Bahamas and Bermuda, in Curran, H.A. & White, B. (eds) *Terrestrial and shallow marine geology of the Bahamas and Bermuda*. Boulder, Colorado: Geological Society of America, Geological Society of America Special Paper 300, p. 251-267.

National Ground Water Association, 1991. *Proceedings of the Third Conference on Hydrogeology, Ecology, Monitoring, and Management of Ground Water in Karst Terranes*. Dublin, Ohio: National Ground Water Association, 793 p.

National Water Well Association, 1986. *Proceedings of the Environmental Problems in Karst Terranes and Their Solutions Conference*. Dublin, Ohio: National Water Well Association, 525 p.

National Water Well Association, 1988. *Proceedings of the Second Conference Environmental Problems in Karst Terranes and Their Solutions*. Dublin, Ohio: National Water Well Association, 441 p.

Neuman, S.P. 1997. Stochastic approach to subsurface flow and transport: A view to the future, in Dagan, G. & Neuman, S.P. (eds) *Subsurface flow and transport: A stochastic approach*. Cambridge: Cambridge University Press, p. 231-241.

Neuman, S.P. & de Marsily, G. 1976. Identification of linear systems response by parametric programming. *Water Resources Research*, 12(2): 253-262.

Palmer, A.N. 1984. Geomorphic interpretation of karst features, in LaFleur, R.G. (ed.) *Groundwater as a geomorphic agent*. Boston: Allen & Unwin, p. 173-209.

Palmer, A.N. 1991. Origin and morphology of limestone caves. *Geological Society of America Bulletin*, 103: 1-21.

Parizek, R.R., White, W.B. & Langmuir, D. 1971. Hydrogeology and geochemistry of folded and faulted rocks of the central Appalachian type and related land use problems. The Pennsylvania State University, Earth and Mineral Sciences Experiment Station – Mineral Conservation Series Circular 82, 213 p.

Parkhurst, D.L., Thorstenson, D.C. & Plummer, L.N. 1980. PHREEQE – A computer program for geochemical calculations: United States Geological Survey, Water-Resources Investigations Report 80-96, 195 p.

Plummer, L.N., Parkhurst, D.L. & Wigley, T.M.L. 1979. Critical review of the kinetics of calcite dissolution and precipitation, in Jenne, E.A. (ed.) Chemical modelling in aqueous systems: Philadelphia, American Chemical Society, *Symposium Series 93*, p. 537-573.

Plummer, L.N., Prestemon, E.C. & Parkhurst, D.L. 1991. An interactive code (NETPATH) for modeling net geochemical reactions along a flow path: United States Geological Survey, Water Resources Investigations Report 91-4078, 227 p.

Plummer, L.N. & Wigley, T.M.L. 1976. The dissolution of calcite in CO_2 saturated solutions at 25°C and 1 atmosphere total pressure. *Geochimica et Comsmochimica Acta*, 40: 191-202.

Pyrak-Nolte, L.J., Nolte, D.D., Myer, L.R. & Cook, N.G.W. 1990. Fluid flow through single fractures, in Barton, N. & Stephansson, O. (eds) *Rock Joints: Proceedings of the International Symposium on Rock Joints*. Rotterdam: A.A. Balkema, p. 405-412.

Quinlan, J.F. 1989. Ground-water monitoring in karst terranes: Recommended protocols and implicit assumptions: United States Environmental Protection Agency, Environmental Monitoring Systems Laboratory, EPA/600/X-89/050, 52 p.

Quinlan, J.F. & Ewers, R.O. 1989. Subsurface drainage in the Mammoth Cave area, in White, W.B. & White, E.L. (eds) *Karst hydrology: Concepts from the Mammoth Cave area*. New York: Van Nostrand Reinhold, p. 65-103.

Raeisi, E. & Karami, G. 1997. Hydrochemographs of Berghan karst spring as indicators of aquifer characteristics: *Journal of Cave and Karst Studies*, 59(3): 112-118.

Rauch, H.W. & White, W.B. 1970. Lithologic controls on the development of solution porosity in carbonate aquifers. *Water Resources Research*, 6: 1175-1192.

Rauch, H.W. & White, W.B. 1977. Dissolution kinetics of carbonate rocks 1. Effects of lithology on dissolution rate. *Water Resources Research*, 13: 381-394.

Shaw, T.R. 1992. History of cave science: The exploration and study of limestone caves, to 1900. Broadway, Australia: Sydney Speleological Society, 338 p.

Shuster, E.T. & White, W.B. 1971. Seasonal fluctuations in the chemistry of limestone springs: A possible means for characterizing carbonate aquifers: *Journal of hydrology*, 14: 93-128.

Smart, C.C. 1983. The hydrology of the Castleguard karst, Columbia Icefields, Alberta, Canada: Arctic and Alpine Research, 15(4): 471-486.

Smart, P.L. & Hobbs, S.L. 1986. Characterisation of carbonate aquifers: A conceptual base, in *Environmental problems in karst terranes and their solutions*, Bowling Green.

Smart, P.L. & Laidlaw, I.M.S. 1977. An evaluation of some fluorescent dyes for water tracing. *Water Resources Research*, 13(1): 15-33.

Stringfield, V.T. 1936. Artesian water in Tertiary limestone in the southeastern states: United States Geological Survey, Professional Paper 517, 226 p.

Sweeting, M.M. 1973. Karst landforms. New York: Columbia University Press, 362 p.

Teutsch, G. & Sauter, M. 1992. Ground water modeling in karst terranes: Scale effects, data acquisition and field validation, in National Ground Water Association (ed.) *Proceedings of the Third Conference on Hydrogeology, Ecology, Monitoring, and Management of Ground Water in Karst Terranes.* Dublin: Ohio, National Water Well Association, p. 17-35.

Thrailkill, J. 1972. Digital computer modelling of limestone groundwater systems: University of Kentucky Water Resources Institute, Research Report No. 50, 71 p.

Thrailkill, J. 1988. Drawdown interval analysis: A method of determining the parameters of shallow conduit flow carbonate aquifers from pumping tests. *Water Resources Research,* 24(8): 1423-1428.

Watson, R.A. & White, W.B. 1985. The history of American theories of cave origin, in Drake, E.T. & Jordan, W.M. (eds) *Geologists and ideas: A history of North American geology.* Boulder: Geological Society of America, Centennial Special Volume 1, p. 109-123.

White, W.B. 1969. Conceptual models for carbonate aquifers: Ground Water, 7(3): 15-21.

White, W.B. 1977a. Conceptual models of carbonate aquifers: Revisited, in Dilamarter, R.R. & Csallany, S.C. (eds) *Hydrologic problems in karst regions.* Bowling Green, Kentucky: Western Kentucky University, p. 176-187.

White, W.B. 1977b. Role of solution kinetics in the development of karst aquifers, in Tolson, J.S. & Doyle, F.L. (eds) Karst Hydrogeology (Memoirs of the 12th Congress of the International Association of Hydrogeologists): Huntsville, Alabama: University of Alabama Huntsville Press, p. 503-517.

White, W.B. 1988. *Geomorphology and hydrology of karst terrains.* Oxford: Oxford University Press, 464 p.

Williams, P.W. 1983. The role of the subcutaneous zone in karst hydrology: *Journal of Hydrology,* 61: 45-67.

Wood, W.W. 1985. Origin of caves and other solution openings in the unsaturated (vadose) zone of carbonate rocks: A model for CO_2 generation. *Geology,* 13(11): 822-824.

Wood, W.W. & Petraitis, M.J. 1984. Origin and distribution of carbon dioxide in the unsaturated zone of the southern high plains: Water Resources Research, 20(9): 1193-1208.

Worthington, S.R.H., Davies, G.J. & Quinlan, J.F. 1992. Geochemistry of springs in temperate carbonate aquifers: Recharge type explains most of the variation, in Chauve, P. & Zwahlen, F. (eds) *Annales scientifique, Mémoire hs n°11 (Actes du 5° colloque hydrologie en pays calcaire),* Universitie Besançon: géologie 2: 341-348.

Quantitative tracer test of the Beech Creek aquifer at the Ammunition Burning Grounds, Naval Surface Warfare Center, Crane, Indiana

STEVEN J. BAEDKE
Department of Geology and Environmental Studies, James Madison University, Harrisonburg, USA

NOEL C. KROTHE
Department of Geological Sciences, Indiana University, Bloomington, USA

ABSTRACT: A quantitative tracer test was conducted on the karst system in the Beech Creek aquifer at the Crane Naval Surface Warfare Center, Indiana. A mixture of 1.6 kilograms of Rhodamine WT (20%) and 18.3 kilograms of Br⁻ ionic tracer were injected into a cased well situated on a karst conduit. Tracers were collected at 14 spring orifices in the study with a combination of passive dye detectors, or passive dye detectors with continuous water sampling and continuous discharge measurements.

Rhodamine WT and Br⁻ were detected at one spring, consisting of a complex of 2 major orifices (Spring A and Spring A') and several diffuse seeps approximately 2000 meters from the injection well. The Br⁻ tracer breakthrough occurred 7 hours after injection, which is one half-hour before Rhodamine detection. The Br⁻ pulse dissipated within approximately 14 hours while Rhodamine WT was still detected days later. Using the time of first arrival for the Br⁻ tracer, the effective conductivity for this system is 286 m hr⁻¹. Approximately 80% of the injected Br⁻ recovered at the 2 major spring orifices and an additional 10-15% of Br⁻ is believed to have issued from the diffuse seeps in the area of the springs that could not be measured for discharge. Calculations suggest that approximately 110% of the injected Rhodamine dye was recovered from the same springs. The near total recovery of the injected tracers from the Spring A complex indicates that this karst system has only one significant outlet. This conclusion is substantiate since dye has not been detected on any of the 14 springs after one year of sampling. The early arrival and short lived Br⁻ tracer pulse, compared to Rhodamine, illustrates the more conservative nature of the ionic tracer.

1 INTRODUCTION

Predicting the pattern of groundwater flow in a karst aquifer is difficult since flow may cross typical flow boundaries defined by the porous media (i.e. groundwater divides), and the direction of flow can change temporally. Tracer tests are generally the most practical and satisfactory method to provide information about the movement of water in a karst aquifer system because of the unique hydrologic characteristics of karst terrains.

Tracing groundwater flow is accomplished by adding a distinctive substance, or tracer, to groundwater and monitoring down-gradient locations of concern. Injection of such distinctive substances, either intentionally or accidentally, has often served to identify point to point connections between input points and resurgences such as springs or pumped wells. If establishing a hydrologic connection between a specific inflow point and a discharge point is all that is desired from the experiment, then simply injecting the tracer and looking for it downstream is all that is needed. This techniques defines a qualitative tracer test. Qualitative tracing techniques use tracers (commonly organic dyes) and passive detectors for tracer recovery to establish approximate flow-routes and groundwater basin boundaries.

To accurately establish water budgets, travel times and flow velocities of the tracer, and/or the amounts of tracer recovered, measurements of the concentration of tracer and discharge at a recovery point need to be measured. This technique defines a quantitative tracer test. Quantitative tracing techniques commonly use tracers (either organic dyes or ionic solutions), automatic samplers taking samples at known time intervals, and continuously recorded discharge measurements to establish parameters of the tracer results. Both qualitative and quantitative tracing techniques have been used in this study.

The injection of the Rhodamine WT constitutes the qualitative portion of the test. The fact that organic dyes can be absorbed on activate charcoal allows for detection of minute quantities of dye in distant areas from the injection point long after injection. Since Rhodamine is absorbed onto soil and rock materials it does not allow for a complete recovery of the dye during the test since as much as 50% of the dye may be absorbed. In this test an ionic tracer (Br$^-$ was injected simultaneously with the Rhodamine since it is not absorbed to soil or rock materials and travels at the velocity of the water in which it is dissolved. Although many qualitative tests have been conducted in karst aquifers few quantitative tests have been conducted or the results published. The advantages of a quantitative test are:

1. The true velocity of flow in the conduit can be ascertained since there is no retardation.

2. Since the ionic tracer is not absorbed to rock or soil materials the percentage of tracer can be determined for each resurgence.

3. If all of the tracer is recovered then doubts about cross-basin flow can be dispelled.

2 HYDROGEOLOGIC SETTING

The study area includes the ABG within the Crane Naval Surface Warfare Center located in southern Indiana. The geology of this area has been well studied by Hunt (1988), Murphy & Ciocco (1990), Murphy (1994), and Barnhill & Ambers (1994). The ABG is within the unglaciated region of the Crawford Upland which is dominated by rugged topography. Rock units observed in the area include Illinois Basin deposits ranging from Mississippian to Pennsylvanian Age (Fig. 1). The West Baden, Stephensport, and Racoon Creek Groups comprise the underlying bedrock. The ammunition burning ground is located at the headwaters of the Little Sulphur Creek drainage valley (Fig. 2). This drainage receives the majority of surface runoff within the ABG and has the potential to spread any contamination from the ABG to off-base locations (Hunt, 1998).

Three aquifers have been identified within the ABG (Hunt, 1988). The upper aquifer is composed of the Golconda/Haney Formation and is underlain by the Indian Springs Shale aquiclude (Fig. 3). This aquifer is present in the western area of the ABG and has been

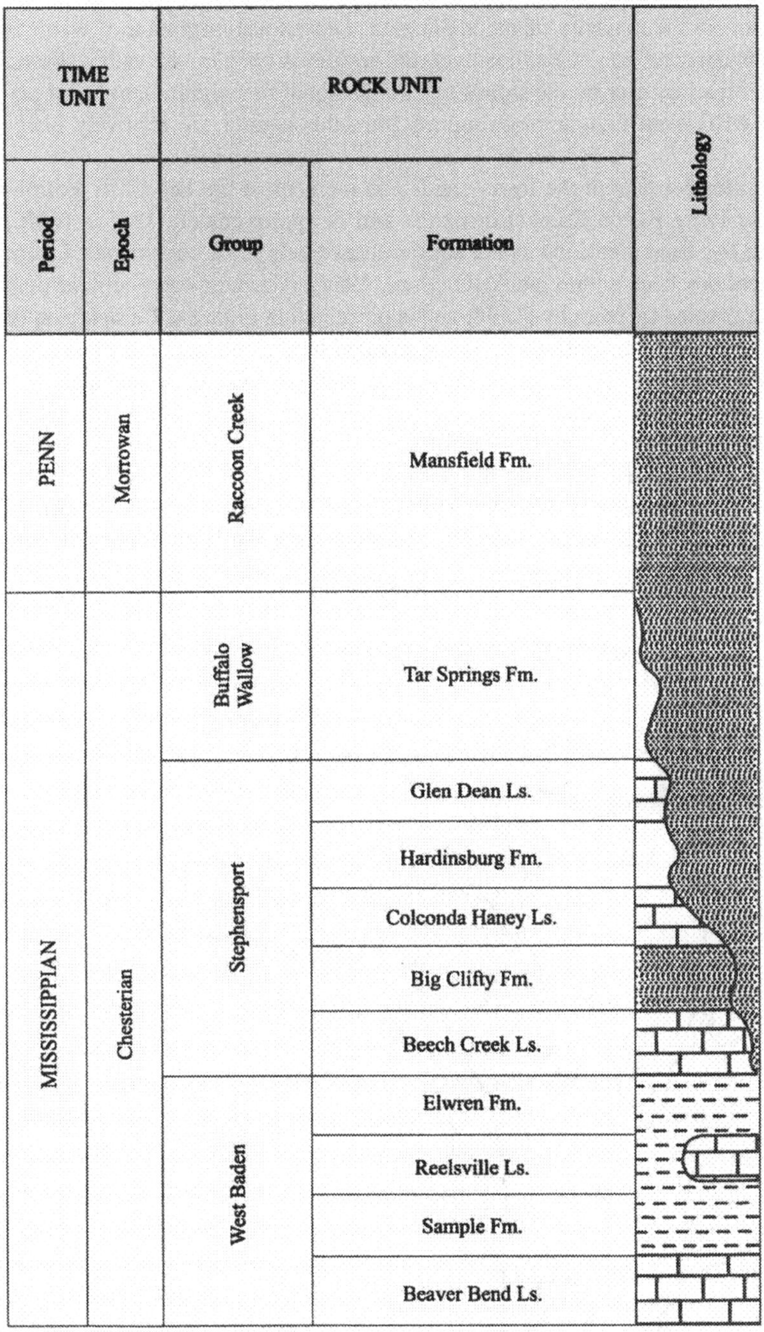

TIME UNIT		ROCK UNIT		Lithology
Period	Epoch	Group	Formation	
PENN	Morrowan	Raccoon Creek	Mansfield Fm.	
MISSISSIPPIAN	Chesterian	Buffalo Wallow	Tar Springs Fm.	
		Stephensport	Glen Dean Ls.	
			Hardinsburg Fm.	
			Colconda Haney Ls.	
			Big Clifty Fm.	
			Beech Creek Ls.	
		West Baden	Elwren Fm.	
			Reelsville Ls.	
			Sample Fm.	
			Beaver Bend Ls.	

Figure 1. Stratigraphic column of rock units within the Naval Surface Warfare Center, Crane Division, Indiana.

removed by erosion in the majority of the ABG area. Downward migration of water is prevented by the underlying aquiclude, however, the aquifer is open on the valley slopes. Water exiting the upper aquifer on the slopes has the potential for migration to lower aquifers within the ABG even though contributions from this aquifer are relatively insignificant.

The middle aquifer consists of the lower sandstone member of the Big Clifty Formation and the under-lying Beech Creek Limestone and is approximately 18.3 m (60ft.) thick. This aquifer has been identified as the aquifer most likely to be contaminated from munitions treatment practices within the ABG (Hunt, 1988). A groundwater contour map of this aquifer was created by Murphy (1994) and is presented in Figure 4. Comparison of Figures 3 and 4 shows that a potentiometric low is created by the collapse in the stratigraphy near the C well complex. Tracers were injected in well O3-CO2-P2 in that area. Recharge to this aquifer occurs at outcrops of the Big Clifty and Beech Creek Formation updip of the ABG (Hunt, 1998), and from infiltration within the ABG where the Indian Springs Shale has been eroded exposing the permeable and highly fractured Big Clifty Sandstone (Hunt, 1988; Barnhill & Ambers, 1994). Solution caverns have developed in the Beech Creek Limestone in areas bordering and underlying the Little Sulphur Creek

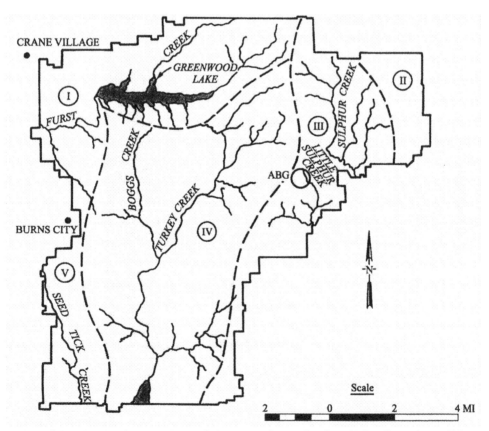

Figure 2. Location of study area and major drainages near the Naval Surface Warfare Center, Crane Division, Indiana (modified from Hunt, 1988).

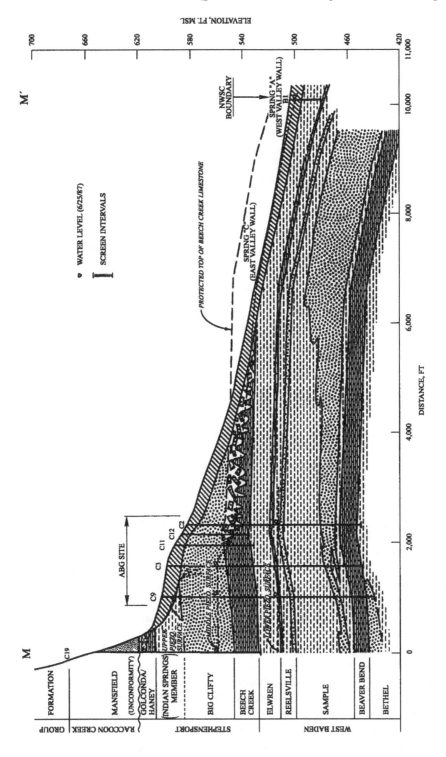

Figure 3. Geologic profile of Little Sulphur Creek Valley, Ammunition Burning Ground (Murphy & Ciocco, 1990).

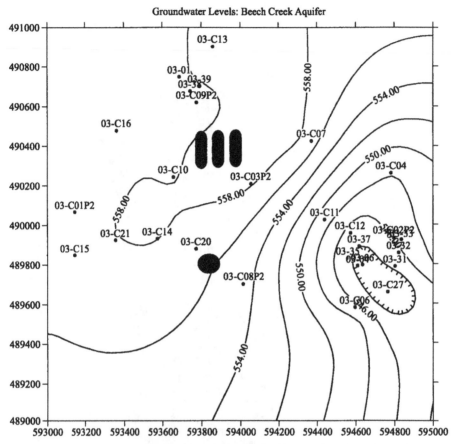

Figure 4. Groundwater contour map of the Ammunition Burning Ground (Dignazio et al., 1998) Elongate ovals ore the current locations of disposal activities. Contour interval is 2 feet and coordinates are in s.

drainage. Water rapidly infiltrates the overlying Big Clifty Sandstone and is forced to flow along limestone joints and fractures (Hunt, 1988). The Beech Creek Limestone matrix has an extremely low permeability, but the overall permeability of the unit is enhanced by the fractures and conduit development (Fig. 3). Therefore, the majority of groundwater within this aquifer is concentrated along these fractures and conduits (Barnhill & Ambers, 1994). Downward flow is prevented at the base of the aquifer by the underlying Elwren Shale aquiclude (Hunt, 1988; Barnhill & Ambers, 1994). The conduits have been shown to be connected to springs discharging in the lower reaches of the Little Sulphur Creek drainage (Murphy & Ciocco, 1990) and have the potential for contaminating off-base locations (Fig. 3).

The lower aquifer is composed of the Beaver Bend Limestone and retains a constant horizontal water level. Recharge to this lower aquifer comes almost entirely from Beaver Bend Limestone outcrops updip to the north of the ABG. The Sample, Reelsville, and Elwren Formations provide 70 ft of aquiclude between the lower Beaver Bend aquifer and the middle Big Clifty/Beech Creek aquifer lending little chance for hydraulic connection between these two aquifers.

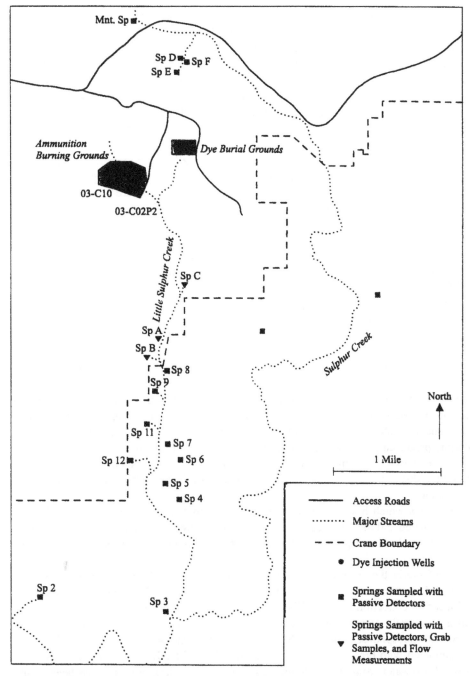

Figure 5. Location of injection wells and location of springs used for monitoring during the tracer test at the Ammunition Burning Grounds.

A hydraulic connection between the Beech Creek aquifer and a complex of springs (Fig. 5) was established by a previous qualitative dye trace at moderate flow conditions at the ABG (Murphy & Ciocco, 1990). During low flow conditions in the karst system, one pound of fluorescein dye was injected into well 03-C02P2 at the ABG. The presence of dye was visually detected between 5 and 24 hours at the Spring A complex (Sp A on Fig. 5), which consists of Spring A and Spring A' and several diffuse seeps. The fluorescein was detected under ultraviolet light from passive detectors placed at the spring orifices. Dye continued to be observed in a 'relatively high' concentration for at least 5 days (Murphy & Ciocco, 1990).

Springs B and C, also in the Little Sulphur Creek valley, were monitored but showed negative or 'weakly positive' indication of tracer. Although dye was not positively identified from Springs B and C, it was hypothesized that these springs are groundwater outlets at high flow conditions. Four springs north of the ABG (Springs E, F, G, and Mountain Spring) were also monitored and showed a 'weakly positive' indication for tracer in 'normal light'. It is, however, possible that the green coloration was caused by algae (Murphy & Ciocco, 1990). Based upon qualitative judgement, it is believed that very little of the total injected dye was recovered during this tracer test (Murphy & Ciocco, 1990).

The results of the previous dye trace have been used to design the quantitative trace experiment for this investigation.

3 DISPOSAL PRACTICES WITHIN THE AMMUNITION BURNING GROUND

The following discussion is summarized from the work of Murphy (1994). The ABG has been in operation since the 1940s and is currently used for the disposal of bare explosives, rocket motors, candles, flares, solvents, red phosphorous, detonators, and fuses. Murphy produced a detailed map showing the locations of all past and present operations within the ABG. A modified version of Murphy's map is presented in Figure 6 which shows the disposal sites as shaded areas with numeric labels that correspond to disposal descriptions.

The areas which have the highest potential for contamination within the ABG have been identified and include area 2, area 6, and the removed ash pile. Area 2 marks the location of 20 clay-lined pans used for the thermal treatment of bulk propellant and high explosives.

Area 6 marks the location of three lagoons which contained liquid sludges from munitions loading processes. These lagoons were modified in 1982 to include liners, covers, and leachate collection systems. The lagoons have been replaced by operations in area 1 and are currently empty. The northern lagoon was used to hold sludge water contaminated with phosphorous compounds. The two central lagoons were used to hold waters contaminated with breakdown components of TNT and RDX from manufacturing processes. Wastewater from explosive manufacturing and loading processes has been identified to have significant amounts of nitrate (Kroschwitz & Howe-Grant, 1991).

In addition, the large circular area designated as Ash Pile in the south of the ABG was used to store ash from thermal treatment operations. The pile included approximately 12,290 lb of ash before it was removed between July 1986 and February 1987. This ash pile was in direct contact with the ground surface for an extended period of time and may have also been a significant source of contamination to the underlying Beech Creek aquifer (DiGnazio et al., 1998).

4 METHODS

The previous dye trace study did not definitively determine whether any spring, except for the Spring A complex, is hydrologically connected to the karst system that was traced. Additionally, since flow measurements were not taken for water emerging from the Spring A complex, it is not possible to quantify the amount of tracer recovered. Unknown tracer recovery is problematic in this situation since it is desirable that the results be able to direct remediation strategies for contaminants in a fracture system. Low tracer recovery could indicate either that significant amounts of tracer are being stored in the fracture system and/or that the tracer is leaving the system at unknown locations.

To improve upon the previous qualitative tracer study, a tracer test was conducted with two objectives: 1. to quantify the amount of tracer recovered from the major springs on NSWC property, and 2. to determine if any springs located on property outside the NSWC received tracer. In order to quantify the amount of tracer recovered, weirs were installed on Springs A', B, and C, and a flume was built for Spring A so that flow measurements could be made (Fig. 5). Transducers with continuous recorders were installed at each spring so that flow could be recorded remotely at frequent intervals. Each weir and flume were calibrated for the range of discharges expected during the tracer test. Calibrations were checked frequently before and during the tracer test and adjusted appropriately when needed.

Field reconnaissance in the study area revealed numerous karst springs issuing from the Beech Creek limestone. Ten additional springs within the Little Sulfur Creek valley and adjacent drainage basins were also monitored for this study (Fig. 5). These springs were monitored to determine if any tracer was moving off of the NWSC base or toward springs to the north of the ABG via conduits, therefore these springs were instrumented with only passive dye detectors (charcoal packets).

Since previous dye tracing in the ABG and activities adjacent to the Little Sulphur Creek drainage basin may produce ambient fluorescence in groundwater samples, a detailed analysis of background fluorescence was conducted to identify the concentrations of specific tracers that should be used during the test. The background analysis showed no detectable amounts of Rhodamine WT (20%) dye in the karst system. Additionally, chemical analysis of the water issuing from the springs showed that Br$^-$ was below the detection limit for the lab equipment. Based on these results, it was determined that Rhodamine WT (20%) would be used as a dye tracer and Br$^-$ would be used as an ionic tracer for the quantitative dye trace.

The amount of Rhodamine WT to be injected was determined by an estimation calculation established by Quinlin (1989) where 1 pound of dye is used per mile of desired trace length. The approximate distance from well 03-C02P2 to Spring 2 (Fig. 3) is 3.5 miles. Therefore it was determined that 3.5 pounds of Rhodamine (1.6 kilograms) were injected. Through consultation with researchers at the Westinghouse Project in Bloomington, IN who had traced with Br$^-$ (McCann, personal communication), it was determined that approximately 18 kilograms of Br$^-$ would be sufficient to trace the springs on

Figure 6. Map of disposal areas at the Ammunition Burning Grounds.

NSWC property. It was calculated that 23.5 kilograms of reagent grade NaBr (solid) dissolved in water dissociates to yield 18.3 kilograms of Br⁻ ion.

To obtain accurate amounts of the dyes detected in the water and charcoals, they were analyzed on a Shimadzu RF 5000 scanning spectrofluorophotometer. This instrument is extremely sensitive and can detect dyes in the part per trillion (ppt) range. A concentration curve was constructed with the dye batch purchased for this test by mixing a 1000 (mg/l) stock solution and diluting to derive concentrations. Bromide was analyzed by ion chromatography on a Dixones 4500 ion analyzer.

5 RESULTS

Well 03-C02P2 was injected with 1.6 kilograms of Rhodamine WT (20%) at 1232 hours (military time) on 5-3-97. This was immediately flushed with a 151 liter solution containing 18.3 kilograms of Br⁻ ion. This was flushed with an additional 19 liters of distilled water. The injection of tracers was completed at 1259 hours.

The karst system was at a high flow stage during the tracer test as 2.5 inches of rain fell the day before the test (5-2-97). The peak flow was topping the V-shaped weirs therefore flow measurements could not be taken. Once the flow receded to where it was flowing through the weirs the trace was initiated. Dye was injected into the wells when measurable discharge at each spring was within the calibrated range of the weirs and flume.

Automatic samplers (ISCO Model 2900 Sampler) at Springs A, A', B, C, and D were immediately set to sample at 30-minute intervals until dye was observed visually discharging from the spring(s). Discharge measurements were taken at Springs A, A', B, and C at 30 minute intervals. All samples were transferred to amber glass bottles, refrigerated, and transported to the lab for refrigeration until analysis. All analyses were completed within a week from the date of sampling. In addition all charcoal packages were changed in springs off the ABG at the initiation of the test.

Dye was visually detected at Spring A and Spring A' at 2000 hours on 5-3-97. A hand sample was taken at 2015. The automatic samplers were reprogrammed to then sample every 15 minutes until 0000 hours on 5-4-97, at which time the sampling interval was changed to every 30 minutes. Charcoal packages were collected weekly for 2 months, biweekly for 4 months then monthly for 12 months after dye injection.

6 DISCUSSION

Rhodamine WT dye was not detected in grab samples or passive detectors ('bugs') from any of the springs except Springs A and A'. When spring discharge, Rhodamine WT, and Br⁻ plotted against time, a breakthrough curve is produced. Breakthrough curves for Spring A and Spring A' sampling sites are shown in Figures 7-10. By analyzing the breakthrough curves, travel time of the tracer can be calculated. For Spring A the breakthrough (first detection) of Rhodamine (Fig. 7) and Br⁻ (Fig. 8) occurred simultaneously at 1930 hours (5-3-97), approximately 7.5 hours after injection. For Spring A', the breakthrough for Br⁻ (Fig. 10) occurred at 1900 hours. The Rhodamine WT (Fig. 9) breakthrough occurred at 1930 hours, one-half hour after the Br⁻ breakthrough for Spring

Tracer
Spring A Rhodamine WT Breakthrough Curve

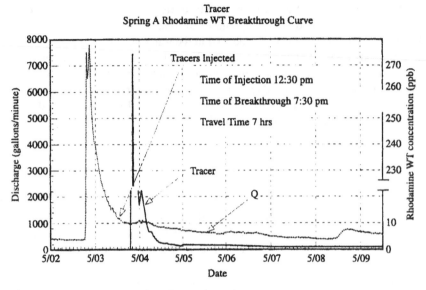

Figure 7. Rhodamine WT breakthrough curve for Spring A.

Spring A Bromide Breakthrough Curve

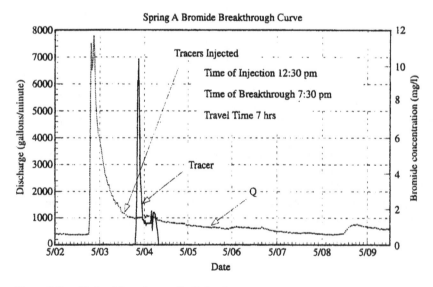

Figure 8. Bromide breakthrough curve for Spring A.

A'. The delay time in the detection of Rhodamine WT is due to the timing of sample collection by the automatic sampler and, based upon the results for Spring A, it is believed that the Rhodamine WT breakthrough occurred very shortly after arrival of the Br⁻ tracer dissipated after about 14 hours while the Rhodamine WT tracer was detected in grab samples for 4 days. The short-lived Br⁻ tracer pulse, compared to Rhodamine, illustrates the more conservative nature of the ionic tracer.

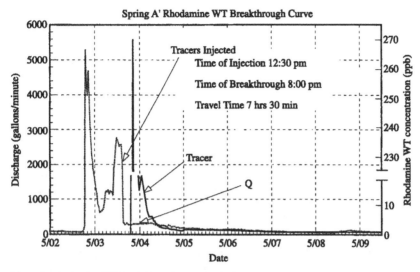

Figure 9. Rhodamine WT breakthrough curve for Spring A.

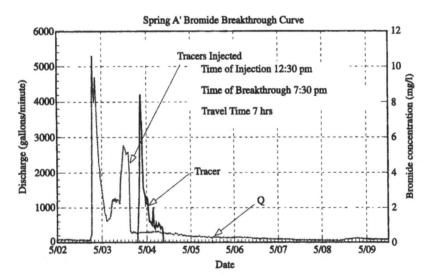

Figure 10. Bromide breakthrough curve for Spring A.

An approximate travel time can be calculated for the conduit from the breakthrough curve data. Br⁻ will be used for these calculations since it appears to have been a more conservative tracer than Rhodamine WT. It is approximately 2000 m from the injection well, 03-C02P2, to the Spring A complex; therefore the hydraulic conductivity is calculated to be 286 m hr⁻¹ number is probably greater considering the possible sinuosity of the conduit. Murphy & Ciocco (1990) estimated the length of the conduit to be 8000 feet based on a sinuosity similar to the valley of Little Sulphur Creek.

The major objective of the quantitative tracer experiment was to determine how much of the tracers could be recovered from the springs located on NSWC property. If most of the dye discharges down Little Sulphur Creek valley at springs on Crane property then remediation is much simpler. To calculate the recovery of tracer, the following equation is used:

$$M = \int_0^\infty QCdt \qquad (1)$$

where M = mass of tracer recovered, Q = discharge, and C = tracer concentration at time t.

Calculations for this test show that 13.9 kilograms of injected Br⁻ (18 kilograms) was recovered, or 77%. During the high flow conditions of the tracer test, when the bromide peak arrived some tracer issued from the seeps between Spring A and Spring A'; therefore some bromide that discharged from the Spring A complex is not accounted for in the previous calculations. It is estimated that the recovery of tracer at the Spring A complex can be increased by an additional 10-15% to account for the discharge with tracer issuing from the seeps on the valley wall. If this is true, approximately 90-95% of the Br⁻ was recovered. Errors in discharge measurements can also account for not recovering the entire mass of the tracer.

The calculation of Rhodamine WT recovery is slightly more complicated than Br⁻ Rhodamine WT (20%) solution as purchased from the distributor is not actually a 20% solution. The stock powder that the 20% solution is made from (at the distributor) only has an average of 86% active ingredient (i.e. Rhodamine WT). Therefore, when this powder is dissolved in a liter of distilled water produces a concentration of 860,000,000 ppb. The supplier then diluted this concentration to a 20% solution and the concentration then becomes 172,000,000 ppb. The quantity of dye that was injected was 1.6 kilograms of the solution containing 172,000,000 ppb. This is only an approximate concentration since, as confirmed by discussions with the distributor of the dye, there is an acceptable range of error of about 10% on the 20% solution (i.e. the 20% solution is actually 172,000,000 ppb ± 17,200,000 ppb).

Calculations show that approximately 3.1×10^8 ppb of Rhodamine WT ($\approx 110\%$) was recovered of the approximately 2.75×10 ppb that was injected. The over estimations is probably due, at least in part, to inexact measurements in preparing the dye. Additionally, since Rhodamine WT was detected for several days after injection, over- or underestimates in discharge measurements and/or Rhodamine WT concentrations during this extended period of time would cause the recovery calculations to be inaccurate. In the case of Br⁻, the recovery calculations would likely be more accurate since the entire breakthrough occurred in 14 hours. Therefore, errors in measuring discharge and/or concentration of tracer would not be made for an extended period of time. In either case, the Rhodamine WT and Br⁻ tracer tests do suggest essentially complete recovery of the tracer from the Spring A complex since springs B and C received no Bromide or Rhodamine WT. Charcoal detectors were collected over weekly intervals at all other springs for two months, bi-weekly for the next four months and then monthly until present. No dye has been detected in any of the other springs including those off-site.

7 CONCLUSIONS

A quantitative tracer test of the karst system in contact with well 03-C02P2 at the ABG was conducted in which 1.6 kilograms of Rhodamine WT (20%) and 18.3 kilograms of Br⁻ tracer were injected into the well. A total of 14 potential discharge points (springs) in the study area (Fig. 5) were monitored with passive dye receptors. Four springs that were believed to be most likely to be hydrologically connected to the injection well were continuously monitored for discharge and continuously sampled for tracer. The remaining 10 springs were monitored with passive dye detectors.

Rhodamine and Br⁻ were detected at only one spring (Figs 7-10), consisting of a complex of two major orifices (Spring A and Spring A') and several diffuse seeps between the springs approximately 2000 meters linear distance from the injection well. The Br⁻ tracer breakthrough occurred one-half hour before the Rhodamine WT dye. The Br⁻ pulse dissipated approximately 14 hours after being detected while Rhodamine WT was still detected days later. Approximately 80% of the injected Br⁻ was recovered at the two major spring orifices. An additional 10-15% of Br⁻ is believed to have issued from the diffuse seeps. Thus approximately 90-95% of the injected Br⁻ can be accounted for. A calculated recovery of 110% of the injected Rhodamine WT dye from the same springs indicates that most of the injected tracers were recovered from the Spring A and Spring A' complex (Fig. 5). This suggests that the karst system that was injected into has only one outlet, which is located at the Spring A and Spring A' complex. This is confirmed since all of the other springs have been monitored for over one year with charcoal detectors with no detection of Rhodamine WT. Using the time of first arrival for the Br⁻ tracer, the travel time of water in the conduit is 286 m hr⁻¹. The other springs do not have any hydrologic connection to the karst conduit system which is confined to the valley of Little Sulphur Creek.

The short-lived Br⁻ tracer pulse, compared to Rhodamine WT, illustrates the more conservative nature of the ionic tracer. Therefore, the calculation of tracer recovery are probably more accurate for the Br⁻ tracer than the dye. Additionally, due to problems with controlling the concentration of dye that is injected into the well, it appears that the recovery calculations for the ionic tracer is more reliable than the dye. However, Rhodamine WT was not significantly retarded in this study and appears to provide a reasonable estimate of travel time in an open conduit system.

Few, if any quantitative tests have been conducted or published concerning karst aquifer systems. Most tests are qualitative with injection a tracer at some point in the flow system and detecting it at a discharge point. The quantitative tracer tests gives you a reasonable estimate of the quantity of tracer recovered in the system. In this case the flow is confined to the major conduit system and discharging at Spring A and Spring A' complex. Charcoals which accumulate the organic dye allows for detection of dye at distances from the injection site that are not practical for ionic traces or quanitative calculations.

This test provides the Crane Naval Surface Warfare Center with information necessary for remediation of the contamination problems. Previous pump and treat tests failed due to the wide range of hydraulic conductivities found in this fractured geologic setting. Results show that remediation should be implemented at the Spring A/A' complex. A quantitative tracer test coupled with a qualitative test can be used in other karst systems to provide information for site remediation.

ACKNOWLEDGMENTS

This research was funded by grants from the Army Corps of Engineers Waterways Experiment Station and the Crane Naval Surface Warfare Center, Indiana. The research was conducted under the supervision of T. Brent and J. Hunsicker of the Environmental Protection Department at Crane. M. Noriega and F. DiGnazio aided in the collection of background data for this study. We are indebted to their help.

REFERENCES

Barnhill, M.L. & Ambers, C.P. 1994. Geology and hydropetrology of the Big Clifty sandstone and Beech Creek limestone aquifer system at the ammunition burning ground, Naval Surface Warfare Center, Crane. Indiana. Geological Survey, Open File Report, 86 p.

Baedke, S.J. 1998. Hydrogeologic and geochemical assessment of the porous media and karstic flow regimes of the Beech Creek Aquifer, Ammunitions Burning Grounds, Naval Surface Warfare Center, Crane, Indiana (Ph.D. Dissertation). Bloomington, Indiana University, 141 p.

Dignazio, F.J., Krothe, N.C., Baedke, S.J. & Spalding, R.F. 1998. $\delta^{15}N$ of nitrate derived from explosive sources in a karst aquifer beneath the Ammunition Burning Ground, Crane Naval Surface Warfare Center, Indiana, USA: *Journal of Hydrology*, 206: 164-175.

Hunt, R.W. 1988. Geology and hydrology of the ammunition burning ground, Crane Naval Weapons Support Center: Department of the Army Waterways Experiment Station, Corps of Engineers Technical Report GL-88-27, 45 p.

Kroschwitz, J.I. & Howe-Grant, M. (eds) 1991. Explosives and propellants. *Encyclopedia of Chemical Technology*, 4th edn. New York: John Wiley, 10: 1-121.

Murphy, W.L. 1994. Final Report, RCRA facility investigation, phase III, groundwater release characterization, SWMU 03/10, ammunition burning ground: US Army Corps of Engineers, Waterways Experiment Station, Vicksburg, Mississippi, prepared for Department of the Navy, Southern Division, Naval Facilities Engineering Command, Charleston, South Carolina, 44 p.

Murphy, W.L. & Ciocco, J. 1990. Dye trace of Beech Creek aquifer; Ammunition Burning Ground, Naval Weapons Support Center Crane, Indiana: US Army Engineer Waterways Experiment Station Final Report, September 1990, prepared for Department of the Navy, Northern Division, Naval Facilities Engineering Command, Philadelphia, Pennsylvania, 76 p.

Quinlan, J.F. 1989. *Ground-water monitoring in karst terranes: recommended protocols and implicit assumptions*. U.S. Environmental Protection Agency, Environmental Monitoring Systems Laboratory, Las Vegas, Nevada, EPA/600/X-89/050, pp. 28-35.

Application of Unit Hydrograph Technique to the Discharge Record at Big Spring, Carter County, Missouri

CAROL M. WICKS & BRIAN BOHM
Department of Geological Sciences, University of Missouri, Columbia, USA

ABSTRACT: In the Ozark region, karst aquifers are an important source of fresh water. To effectively manage the water supply, a predictive model of groundwater flow and solute transport is necessary. The linear systems approach holds potential for use as a predictive model. One drawback to this approach is the lack of an easy-to-use method to derive a transfer function that can be used to model the response to individual storm events, multiple-day events, and year-long series. In this study, we determine the transfer function relating excess precipitation to discharge from Big Spring, Carter County, Missouri by using unit-hydrograph technique. We calculated the discharge from Big Spring over a range of hydrologic conditions and evaluated the range of applicability of the unit hydrograph technique.

We found that the unit hydrograph technique provided a simple and quick method of obtaining the transfer function of a large spring basin to recharge events without the inherent problems of deconvolution and without concerns about errors in the input series. The unit hydrograph technique was successfully used to calculate the discharge from Big Spring generated by large isolated storms, multiple-day events, and a year-long series. However, for a hydrologically open basin, such as Big Spring, the use of a scaling factor was necessary to obtain a closed water budget. Therefore, true prediction of the discharge was not possible. A better means of determining the temporal and spatial distribution of recharge to the groundwater basin is needed before a closed water balance can be obtained and the scaling factor eliminated.

1 INTRODUCTION

In the Ozark region of Missouri and Arkansas, karst aquifers are an important source of fresh water. To effectively manage the water supply in this region, a predictive model of groundwater flow and contaminant transport through these aquifers is necessary (Back & Herman, 1997). Karst aquifers have solutionally enlarged joints and conduits that are difficult to characterize hydraulically and that act as pathways for rapid movement of groundwater and contaminants. Our inability to characterize the hydraulic properties of large karst basins is the primary reason that few predictive models of groundwater flow and solute transport through karst aquifers have been developed.

One approach that has proved useful is the linear systems approach (Dreiss, 1982; Knisel, 1972; Estrela & Sahuquillo, 1997). One advantage of the linear systems approach is that it does not rely on the specification of the size, shape, or location of the internal structure of the aquifer. The approach allows one to obtain a transfer function (unit hydrograph, linear transform, kernel) that can be used to predict the response of a spring basin to varying hydrologic conditions. The linear systems approach treats the basin as a 'black box' that characteristically operates on input functions to produce an output response (Dreiss, 1982) and is represented by:

$$y(t) = \int_{t=0}^{\infty} h(t-\tau)x(\tau)d\tau \tag{1}$$

where $y(t)$ is the output series, $x(\tau)$ is the input series, and $h(t-\tau)$ is a transfer function. This approach was first derived for hydrologically closed surface water basins for which the output series was discharge and the input series was excess recharge (precipitation minus changes to soil moisture and evapotranspiration). In karst basins, the output series is discharge from a spring or springs. The input series is recharge to the groundwater system. That recharge can occur through infiltration through the unsaturated zone and seepage through sinkholes (autogenic recharge) and through losing streams (allogenic recharge) (White & others 1995).

Several researchers (Blank & others, 1971; Neuman & de Marsily, 1976; Dreiss, 1982, 1989) have shown that derivation of the transfer function by the method of deconvolution is a mathematically ill-posed problem that results in unstable (oscillatory) functions. Estrela & Sahuquillo (1997) have used the relationship between the slopes of the line segments along the recessional limbs (observed when plotted in log Q against time) to derive the values of the transfer function. Estrela & Sahuquillo (1997) did not use deconvolution, however, their method accounted of the response of basins to a year-long recessional not to isolated recharge events. Whereas, the linear systems approach holds potential for use as a predictive model, one drawback is the lack of an easy-to-use method to obtain a transfer function that can be used to model the response to individual storm events, multiple-day events, and year-long series.

In this study, we evaluate the use of unit hydrograph technique (Dingman, 1994) to derive the transfer function for a karst basin. We determined dimensionless unit hydrographs for Big Spring basin (Dingman, 1994) and compared the dimensionless unit hydrographs to kernels that were derived by deconvolution by Dreiss (1982). Using one of the six dimensionless unit hydrographs, we calculated the discharge from Big Spring over a range of hydrologic conditions and evaluated the range of applicability of the unit hydrograph technique.

2 DESCRIPTION OF THE STUDY AREA

The study area is the Big Spring basin that includes parts of Carter, Shannon, Howell, and Oregon Counties, Missouri. Big Spring is located at NW ¼ NE ¼ section 6, T. 26 N., R. 1 E. on the Poplar Bluff 30' × 60' quadrangle, 36°57'05" N latitude, 90°59'36" W longitude (Fig. 1). The region in which Big Spring lies shows typical karst features of the Ozark region including sinkholes, caves, and large springs (Vineyard & Feder, 1982). Big

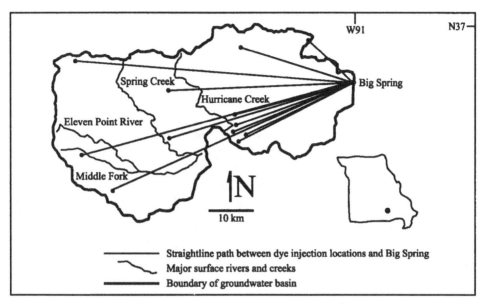

Figure 1. Map of the Big Spring groundwater basis including an index map of the state of Missouri.

Spring, one of the largest springs in the Ozark region, is located in Mark Twain National Forest (Vineyard & Feder, 1982). Daily discharge records have been collected from Big Spring from 1921 through September 1996 by the US Geological Survey (http://h2o-nwisw.er.usgs.gov/nwis-w/MO/?statnum = 07067500). The mean flow of the spring for the period of record is 12.7 m^3/s. The minimum flow was 6.68 m^3/s and the maximum estimated flow was 36.8 m^3/s (Vineyard & Feder, 1982). The maximum flow from the spring is estimated because the nearby Current River back-floods the spring reach during high-flow events.

The climate of the region is temperate and humid (http://www.ncdc.noaa.gov/ol/ climate/climatedata.html#DAILY). The average annual precipitation of 113 cm is distributed mostly in two rainy seasons, the spring and fall. Winter precipitation occurs as rainfall with a trace amount of snow. Summer thunderstorms are localized and occasionally produce significant quantities of rainfall. The average annual temperature is 24.4°C with extreme temperatures of –0.6°C in the winter and 38.0°C in the summer.

The Gasconade, Eminence, and Potosi Formations, which were first karstified during the late Cambrian and Ordovician (Imes & Emmett, 1994), compose the Ozark Plateau aquifer (Fig. 2). The Potosi Formation produces the greatest well yields, which vary over orders of magnitude (Robertson, 1963). These formations are nearly horizontal, massively bedded dolomites of Cambrian-Ordovician age. All three formations are exposed within the Big Spring basin. Big Spring issues from the Eminence Dolomite (Dreiss, 1982).

The 1750-km^2 recharge area for Big Spring was defined based on dye tracing (Aley, 1978) and topography (USGS, 1983, 1984, 1986a and b). The Eleven-Point River and its tributaries, Spring Creek and Hurricane Creek, drain across the boundary of the Big Spring recharge zone in the western portion of the basin (Fig. 1). The upgradient section of the basin is drained by streams and rivers that lose water to the groundwater reservoir

(Aley, 1978). The downgradient section of the basin is internally drained through sink-holes and exhibits no integrated surface-water drainage patterns.

3 METHODS

3.1 *Obtaining the transforms*

Dreiss (1982) derived by deconvolution kernels from six discrete storm events that oc-curred between 1965 and 1972 over the Big Spring basin. Each kernel was digitized from Figure 13 in Dreiss (1982). The data were digitized at daily intervals as the precipitation and discharge data from which the kernels were derived were based on daily data. For the same six storms events, dimensionless unit hydrographs were derived following the pro-cedure outlined by Dingman (1994) and as shown in Table 1. The correlation coefficient between the daily values of the unit hydrograph and the daily values of the kernel for each storm was calculated and tested for significance at the $\forall = 0.05$ level (H_o: $r = 1$; H_a: $r \neq 1$) (Blank, 1980).

3.2 *Calculation of the discharge*

For isolated storm events, for multiple-day recharge events, and for a year-long series, the discharge from the spring was calculated using the dimensionless unit hydrograph that was obtained from the July 1972 data. The calculated discharge records were compared to the observed discharges by calculating the correlation coefficient between the data sets. The correlation coefficients (r) were tested for significance (H_o: $r = 0$; H_a: $r \neq 0$; $\forall = 0.05$) (Blank, 1980).

Each daily value of the transfer function was multiplied by the amount of excess pre-cipitation delivered to the basin (EP, described below), the area of the basin (A), and a scaling factor (SF, also described below). The product of those four terms was summed over the duration of the event (Equation 2):

$$Q_{calculated,\,eventflow} = \sum h(t - \tau) * EP * A * SF * d\tau. \tag{2}$$

Baseflow is added to the calculated event flow giving the calculated discharge.

Daily precipitation and temperature records from six National Weather Service sta-tions were obtained from the State Climatology Office (S. Hu, pers. comm., 1996). For each recharge event, the amount of excess precipitation (EP) delivered to each station was calculated using a moisture-balance method (Thornthwaite & Mather, 1957) and us-ing Theissen polygons (there is no orographic effect over the basin). In order to establish the initial soil moisture conditions for the moisture-balance method, a year-long series of data (1970) was used in the moisture-balance method. The calculated soil moisture con-dition at the end of year 1970 was carried forward as the initial soil moisture condition for the 1971 water year. Once started, moisture balance calculations were carried forward for all subsequent water years. A 7.6 cm soil moisture storage capacity that was used in the derivation of the kernel (Dreiss, 1982) was used in all the calculations. The areally-averaged excess precipitation was used as the amount of water delivered to the basin during a recharge event.

Table 1. Derivation of the dimensionless unit hydrograph.

Date	Discharge (m³/s)	Days	Baseflow (m³/s)	Storm flow (m³/s)	Unit hydrograph
07.27.1972	327	0	327	0	0.00
07.28.1972	351	1	327	24	0.07
07.29.1972	385	2	327	58	0.17
07.30.1972	382	3	327	55	0.16
07.31.1972	368	4	327	41	0.11
08.01.1972	361	5	328	33	0.10
08.02.1972	355	6	328	27	0.08
08.03.1972	349	7	328	21	0.06
08.04.1972	345	8	328	17	0.05
08.05.1972	342	9	328	14	0.04
08.06.1972	341	10	328	13	0.04
08.07.1972	339	11	328	11	0.03
08.08.1972	338	12	328	10	0.03
08.09.1972	335	13	329	6	0.02
08.10.1972	332	14	329	3	0.01
08.11.1972	332	15	329	3	0.01
08.12.1972	332	16	329	3	0.01
08.13.1972	329	17	329	0	0.00
Sums				339	1.00

For each and every recharge event, the volume of water that was discharged from the spring was calculated. For each and every recharge event, the amount of water delivered to the basin (EP) was adjusted to equal the amount that was discharged from the spring by use of a scaling factor (SF). As more water was delivered to the basin than was discharged from the spring, the scaling factors (one per recharge event) were less than 1 (Table 1).

4 RESULTS

4.1 *The transforms*

Unit hydrographs were derived for five of the six storms events used by Dreiss (1982). For one event (July 1968), we could not find a storm event. For the other five events, the correlation coefficients between the daily values of the dimensionless unit hydrograph and those of the digitized kernel ranged from 0.93 to 0.99 (Fig. 2).

4.2 *The discharge records*

For the recharge events of interest, the observed discharge and the calculated discharge are shown in Figures 3, 4, and 5. The maximum discharge, the time at which the maximum discharge occurred, and the duration of the event for both the observed and calculated data are shown in Table 2. The scaling factors and the correlation coefficients between the calculated and observed data are shown in Table 2.

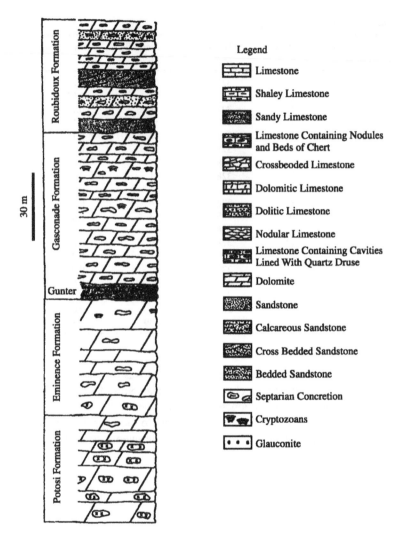

Figure 2. Stratigraphic column of the study area modified from Hayes and others (1961) and Martin and others (1961).

5 DISCUSSION

5.1 *Comparison between the kernels and the unit hydrographs*

Unit hydrograph method of derivation of the transfer function is easy to use and uses only the discharge data. The dimensionless unit hydrograph is not unstable as a result of errors in the input series because the input series is not used in the derivation process. The match between the kernels and the dimensionless unit hydrographs is very good ($r > 0.93$, Fig. 3). Thus, unit hydrograph technique can be used to obtain a transfer function that relates excess precipitation to discharge for Big Spring basin.

Table 2. Results of the calculated and observed discharge events.

Event		Maximum discharge (m^3/s)	Days to maximum discharge (days)	Duration (days)	Scaling factor	r
July 1972	Observed	10.9	3	17	0.0129	0.98
	Calculated	10.8	2			
November 1992	Observed	15.9	4	20	0.0307	0.97
	Calculated	17.4	4			
September 1975	Observed	14.3	3	24	0.0284	0.94
	Calculated	15.6	4			
November 1993	Observed	25.2	5	31	0.2236	0.84
	Calculated	29.8	7			
December 1973	Observed	19.8	2	29	0.0542	0.98
		21.2	7			
	Calculated	20.8	3			
		20.1	7			
June 1973	Observed	25.5	3	25	0.0871	0.94
		26.7	5			
		20.8	12			
	Calculated	25.0	3			
		19.4	13			
October 1991	Observed	18.4	4	51	0.0350	0.87
		17.8	22-23			
		17.0	36			
	Calculated	18.1	5			
		20.4	24			
		17.8	36			
February 1994	Observed	18.2	4	44	0.0387	0.66
		18.1	21-22			
		16.6	36-37			
	Calculated	20.0	5			
		20.9	19			

5.2 *The effect of parameters on the calculations*

The effect that each of the five components on the right-hand-side of Equation (2) has on the calculated discharge needs to be described. The components are the time step, the scaling factor, area of the groundwater basin, the excess precipitation, and the transfer function. The effect that the area of the basin, that the excess precipitation, and that the shape of the transfer function (kernel) have on the predictive ability of the method have been described by Dreiss (1982). We describe the effect of the time step and the scaling factor on the predictive ability of the method.

The scaling factor was used to close the water balance between the input into the groundwater basin (EP × A) and the output from the spring (Q) during recharge events. The amount of water that discharges from Big Spring comes from both allogenic and autogenic sources, the proportion of water from either at any time is unknown, thus, it was not possible to determine a consistent scaling factor for all storm events. In karst basins that are hydrologically open, such as Big Spring, the use of a scaling factor is necessary to force the water budget to close. In this case, an overestimation of the discharge in any portion of the recession results in an underestimation along some other portion of the recession. Hoke (1998) found that for Maramec Spring Basin, which is hydrologically closed, the use of a scaling factor is unnecessary.

The time step used in this study was one day. All data were daily measurements, and all calculations were performed to determine daily discharges. Because we used daily data, we cannot differentiate between a 6-cm rainfall event that occurs uniformly over 24-hours and a 6-cm rainfall event that occurs uniformly over 3-hours. Our ability to resolve the input function (rate of recharge to the groundwater basin), the output response (discharge from the basin), and the transfer function is limited to twice the time step or two days. Having all data sets collected at more frequent intervals would improve our ability to resolve the transfer functions, discharge records, and input functions.

5.3 *The calculated and observed discharge records*

The transfer function was derived from the July 1972 event, which produced a maximum discharge of 10.8 m³/s and had a 17-day duration. The first test was to reproduce the storm from which the transfer function was derived. The correlation coefficient between the calculated discharge and the observed discharge was 0.98 (Table 2 and Fig. 4). The transfer function was applied to three other single-peak events, September 1975, November 1992, and November 1993. The match between the calculated discharges and the observations was good ($0.84 < r < 0.98$). The transfer function was used to calculate and reproduce the discharge from storm events with 2.5 times the maximum discharge and a duration of nearly twice as long.

The next predictions were based on recharge events (December 1973, June 1973, October 1991, and February 1994) that caused multiple peaks within the discharge record (Table 2 and Fig. 5). For the December 1973 event, the calculated discharge was greater than that observed up to day 6 but less than that observed for the second peak and along the majority of the recessional limb. The calculated time to the first peak was lagged by

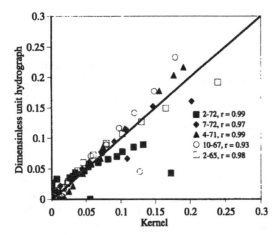

Figure 3. Comparison between the dimensionless unit hydrographs and kernels that were derived for the same storms. The solid line is the line along which the value of the dimensionless unit hydrograph equals that of the kernel. The legend includes the month and year of the recharge event and the correlation coefficient between the calculated and observed discharge records.

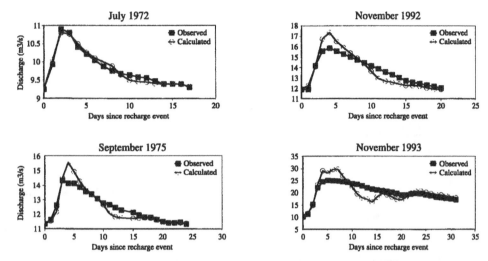

Figure 4. Figures of the calculated and observed discharge records caused by isolated storm events in a) July 1972, b) November 1992, c) September 1975, and d) November 1993. Scales of axes are varied.

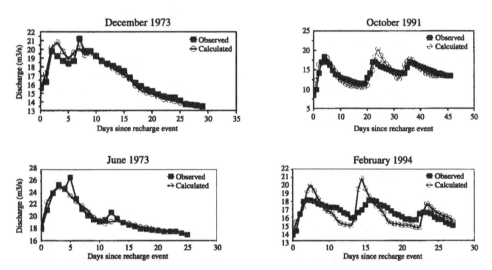

Figure 5. Figures of the calculated and observed discharge records caused by multiple-day events in a) December 1973, b) June 1973, c) October 1991, and d) February 1994. Scales axes are varied.

one day, however the calculated time to the second peak matched the observed time. The match between the calculated and the observed discharge indicate good agreement ($r = 0.98$). For the June 1973 multiple peak event, the transfer function smoothed two peaks into a single peak and the calculated maximum discharge was less than that observed ($r = 0.94$). For the October 1991 event, the first and third maximum discharges were matched both in timing and magnitude. The first and third recessional curves were matched well also. However, the second peak was not matched well in terms of timing or magnitude ($r = 0.87$). For the multiple-peak event that occurred in February 1994, the

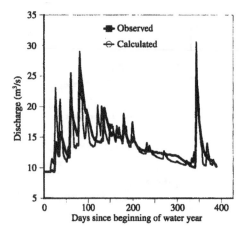

Figure 6. Figure of the calculated and observed discharge records for the water year 1993 and part of 1994.

calculated maximum discharge was greater than the observed maximum discharge, the recessional portions of the curves were underestimated, however, the timing of the peak discharges was good ($r = 0.66$).

The final prediction was for year-long series of data the water year of 1993 and a portion of the water year of 1994. The calculated discharges were greater than the observed maximum and less than those observed along the recessional portion of the curves (Table 2 and Fig. 6). The fit between the calculation based on the transfer function and the observational data was 0.66. However, the calculated discharge did not accurately reproduce the recessional response of spring and it produced false peaks during the late spring and summer. The false peaks were because the calculated excess precipitation resulted in a 'storm' event, whereas, natural conditions did not result in a storm event.

6 CONCLUSIONS

The dimensionless unit hydrographs and kernel functions were derived from the observed increase in discharge from Big Spring as a result of isolated storms that occurred over the basin. There is no difference between the transfer functions (Fig. 3). A representative transfer function was used to calculate discharge records caused by larger isolated storms, by storms that resulted in multiple-peak discharge events, and to a year-long record with reasonable success (Table 2). Thus, the dimensionless unit hydrograph technique provides a simple and quick method of obtaining the transfer function of large spring basins to recharge events without the inherent problems of deconvolution and without concerns about errors in the input series.

For a hydrologically open basin, such as Big Spring, the use of a scaling factor is necessary to obtain a closed water budget. Therefore, a true prediction of the discharge was not possible. However, the method did produce a match for the time to occurrence of the maximum discharge for both single and multiple peak events. Errors occurred where the maximum discharge was overestimated and the discharge along the recessional limb was underestimated. Because the volume of water added to the basin (EP × A × SF) was forced to equal the volume of water discharged from the basin (the water budget was closed), an overestimation in one portion of the curve forced an underestimation in another. A better means of determining the amount of water that actually recharges these large karst basins is needed before a single value of the scaling factor could be determined.

ACKNOWLEDGEMENTS

We thank Ralph Davis and William Jones for providing helpful comments on the manuscript and the Research Board of the University of Missouri for supporting this research project.

REFERENCES

Aley, T. 1978. A predictive hydrologic model for evaluating the effects of land use and management on the quantity and quality of water from Ozark Springs. *Missouri Speleology*, 18: 185 p.

Back, W. & Herman, J.S. 1997. American hydrogeology at the millenium: An annotated chronology of 100 most influential papers. *Hydrogeology Journal*, 5: 37-50.

Blank, D., Delleur, J.W. & Giorgini, A. 1971. Oscillatory kernel functions in linear hydrologic models. *Water Resources Research*, 7: 1102-1117.

Blank, L. 1980. *Statistical procedures for engineering, management, and science.* McGraw-Hill Book Company, New York, 649 p.

Dingman, S. L. 1994. *Physical hydrology.* New York: Macmillan Publishing Company, 575 p.

Dreiss, S.J. 1982. Linear kernels for karst aquifers. *Water Resources Research*, 18: 865-876.

Dreiss, S.J. 1989. Regional scale transport in a karst aquifer: 1. Component separation of spring flow hydrographs. *Water Resources Research*, 25: 117-125.

Estrela, T. & Sahuquillo, A. 1997. Modeling the response of a karstic spring at Arteta Aquifer in Spain. *Groundwater*, 35: 18-24.

Hayes, W.C. & Knight, R.D. 1961. Cambrian system, in Koenig, J.W. (ed.) *The stratigraphic succession in Missouri, v. XL,* second series. Rolla, Missouri, Division of Geology and Land Survey, p. 14-20.

Hoke, J.A. 1998. Modeling groundwater flow and solute transport. Maramec Spring Basin (MS Thesis). Columbia, University of Missouri, 68 p.

Imes, J.L. & Emmett, L.F. 1994. Geohydrology of the Ozark Plateau Aquifer System in parts of Missouri, Arkansas, Oklahoma, and Kansas: U.S. Geological Survey Professional Paper 1414-D, 127 p.

Knisel, W.G. 1972. Response of karst aquifers to recharge. Colorado State University *Hydrology Papers*, no. 60, 48 p.

Martin, J.A., Knight, R.D. & Hayes, W.D. 1961. Ordovician system in Koenig, J.W. (ed.) *The stratigraphic succession in Missouri*, v. XL, second series. Rolla, Missouri, Division of Geology and Land Survey, p. 20-31.

Neuman, S.P. & de Marsily, G. 1976. Identification of linear systems response by parametric programming: *Water Resources Research*, 12: 253-262.

Robertson, C.E. 1963. Water well yield map of Missouri: Missouri Division of Geology and Land Survey, scale 1:750000, 1 sheet.

Thornthwaite, C.W. & Mather, J.R. 1957. *Instructions and tables for computing potential evapotranspiration and the water balance.* Drexel Institute of Technology, Laboratory of Climatology, Publications in Climatology, 10: 464 pp.

US Geological Survey, 1983. Poplar Bluff 30 × 60 Topographic Map. U.S. Geological Survey, 36090-E1-TM-100, scale 1:100,000, 1 sheet.

US Geological Survey, 1984. West Plains 30 × 60 Topographic Map. U.S. Geological Survey, 36091-E1-TM-100, scale 1:100,000, 1 sheet.

US Geological Survey, 1986a. Piedmont 30 × 60 Topographic Map. U.S. Geological Survey, 37090-A1-TM-100, scale 1:100,000, 1 sheet.

US Geological Survey, 1986b, Spring Valley 30 × 60 Topographic Map. U.S. Geological Survey, 37091-A1-TM-100, scale 1:100,000, 1 sheet.

Vineyard, J.D. & Feder, G.L. 1982. Springs of Missouri. Missouri Geological Survey and Water Resources, *Water Resources Report 29*, 267 p.

White, W.B., Culver, D.C., Herman, J.S., Kane, T.C. & Mylroie, J.E. 1995. Karst lands, *American Scientist,* 83: 450-459.

The use of regression analysis to predict nitrate-nitrogen concentrations in springs of northwest Arkansas

ERIC W. PETERSON
University of Missouri, Department of Geology, 101 Geological Sciences, Columbia, USA

RALPH K. DAVIS & J. V. BRAHANA
University of Arkansas, Department of Geology, Fayetteville, USA

ABSTRACT: Five springs in northwest Arkansas have been monitored for non-point source nitrate contamination from 1995 to 1998. Nitrate-N concentrations in spring water samples range from 2.5 mg/L to 15 mg/L. Northwest Arkansas is a karstified region and is a primary poultry-producing area in the United States. Poultry waste is the main source of nitrate in the springs.

The springs were sampled at discrete times over the hydrograph of selected storm events to assess water quality. Additionally, the springs were equipped with data loggers, which recorded stage and specific conductance at 15-minute intervals. Unfortunately, there were no continuous data available concerning nitrate as nitrogen (NO_3-N) concentrations. However, graphical interpretation of the data revealed a possible connection between the nitrate concentrations and two continuously monitored attributes, stage and specific conductance. The goal of this work was to determine if a relationship did exist among NO_3-N concentrations, stage and specific conductance. If a relationship did exist, a secondary goal was to apply the derived relationship to known stage and specific conductance data to simulate NO_3-N concentrations.

Regression analysis was used to determine if a relationship among nitrate concentration and both stage and specific conductance existed. A number of regression analysis techniques were utilized to determine the 'best' approximating equation. First, non-linear regression analysis was performed using the stage as the independent variable. Then multiple regression analysis was conducted followed by a piecewise multiple regression analysis. All multiple regression techniques used stage and specific conductance as the independent variables.

Using a 95 percent confidence level for the regression analysis produced correlation coefficients (r^2) which ranged from 0.12 to 0.99, with 10 of 15 coefficients greater than 0.60. The resulting regression equations were used to simulate nitrate concentrations. These simulated nitrate concentrations were compared to the measured nitrate concentrations produced over the storm hydrograph. Average error values between the measured and simulated NO_3-N concentrations were under 15%. The regression equations were then used to predict the nitrate concentrations at all points where stage and specific conductance were known.

The springs have different proportions of conduit to diffuse flow; therefore, for each spring the generated equations were different. The regression analysis showed the relationship among the stage, specific conductance, and nitrate concentrations varied at each spring. However, the analysis does provide a means to approximate nitrate concentrations when limited data are available. This technique could be applied to other karst areas subject to significant animal waste application.

1 INTRODUCTION

Northwest Arkansas is underlain predominantly by carbonate rocks that have been weathered and dissolved to form a mantled karst landscape. Limestone dissolution has created near-surface karst features that are not readily visible throughout much of northwest Arkansas, but are reflected in the hydrogeologic characteristics including losing streams, numerous seeps and springs, clay-filled fractures and bedding planes, and rapid ground-water flow rate (Leidy & Morris, 1990; Orndorff et al., 1997; Peterson, 1998; Coughlin, 1975). As a result of the karst features, the hydrology of northwest Arkansas is dominated by underdraining, rapid flow, heterogeneous, anisotropic flow paths.

A major concern in northwest Arkansas is the risk associated with poultry waste contamination of ground water in these shallow karst aquifers. Nationally, Arkansas is ranked first in broiler (chicken) production, third in turkey production, and eighth in laying hens (Arkansas Agricultural Statistics Service, 1995). The annual waste production associated with each bird is estimated at 1.13 kilograms (kg) (Scott, 1989) and the nitrogen contribution per bird is 0.4 kg (Goldberg, 1970). With 1.1 billion broilers, 26 million turkeys, and 16 million laying hens in production annually in Arkansas (Arkansas Agricultural Statistics Service, 1995), there are approximately 1.3 million metric tons of waste and a nitrogen contribution of 0.4 million metric tons requiring disposal per year. In addition to nitrogen, animal waste contains two other major nutrients, phosphorus and potassium, essential for plant production (Davis et al., 1995). Therefore, animal waste is used as a source of fertilizer for pastures.

Since 1960, fertilizer use in northwest Arkansas has increased substantially as the poultry industry has grown. Fertilizer-use estimates show that the application of nitrogen fertilizers has increased 152 percent from 1965 to 1985 (Alexander & Smith, 1990). Because of the non-point source contamination resulting from fertilizer application, springs in northwest Arkansas have shown elevated concentrations of nitrates as nitrogen (NO_3-N). Background NO_3-N concentrations for spring water have been reported as well below 1 milligram per liter (mg/L); however, current baseflow concentrations range between 3 to 13 mg/L NO_3-N (Peterson, 1998; Orndorff et al., 1997; Adamski & Steele, 1988; Leidy & Morris, 1990). Despite the majority of NO_3-N concentrations being below the United States Environmental Protection Agency maximum contaminant level (MCL) of 10 mg/L, public concerns over possible health risks have been raised.

Although selected springs are sampled periodically during the year, continuous NO_3-N concentrations are not available for the springs. Acquisition of continuous concentration data would be a useful monitor of NO_3-N contamination. Additionally, continuous concentration data could provide load estimates over storm hydrographs or during baseflow conditions, and the temporal variation of NO_3-N concentrations may provide insight into the dominant processes affecting transport and fate of NO_3-N in these and similar hydrologic environments.

Graphical comparison of stage and NO_3-N concentrations illustrated that as stage increased, usually NO_3-N concentrations also increased. Another result of the graphical interpretation of the data was that as specific conductance (SP) decreased, the NO_3-N concentrations generally decreased. Based on visual observations of spring stage, NO_3-N concentration curves, and SP over a storm hydrograph, a relationship among the three parameters appeared to exist. This study was designed to quantify this relationship and utilize the relationship to predict NO_3-N concentrations over an entire hydrograph.

The goals of this study were:

1. Define the relationships among NO_3-N concentrations, stage, and SP at any given point in spring stage.

2. Extrapolate the relationship among NO_3-N concentrations, stage, and SP over storm hydrographs.

3. Determine whether a relationship developed for one spring system can be used to simulate NO_3-N concentrations at another spring system.

4. Identify hydrogeologic similarities and differences among springs.

2 HYDROGEOLOGY

The study focuses on five springs in northwest Arkansas: Decatur Spring, Braly Spring, Little Wildcat Spring, Tanyard Spring, and Stafford Spring. The locations of the five springs are presented in Figure 1. Three of the springs, Decatur Spring, Little Wildcat Spring, and Stafford Spring, discharge from the Boone Formation. Braly Spring discharges from the Pitkin Formation, and Tanyard Spring discharges from the Batesville Sandstone. Conceptual models (Fig. 2) illustrate the stratigraphic and lithologic settings of the five springs.

The Pitkin Formation is a fossiliferous, oolitic, bluish-gray limestone (Croneis, 1930). The formation is significantly karstic; however, flow through the Braly Spring basin is classified as diffuse with a small conduit component (Orndorff et al., 1997). The Pitkin

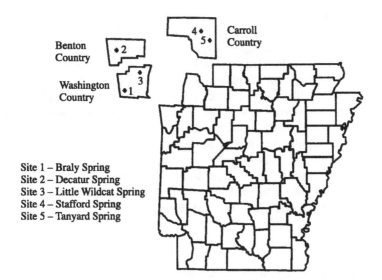

Site 1 – Braly Spring
Site 2 – Decatur Spring
Site 3 – Little Wildcat Spring
Site 4 – Stafford Spring
Site 5 – Tanyard Spring

Figure 1. Location of sampled springs in northwest Arkansas.

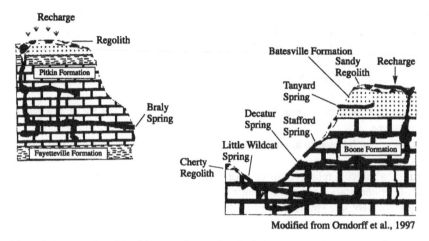

Modified from Orndorff et al., 1997

Figure 2. Conceptual models of the ground-water flow in the karst and non-karst systems for the five springs.

Formation conformably overlies the Fayetteville Shale, which conformably overlies the Batesville Formation.

The Batesville Formation is dominated by a thin-bedded, light-yellow, buff or brown, quartz sandstone. The sandstone is medium grained, but may be coarse- or fine-grained. Typically, the sand is well-cemented with calcite (Branner, 1940; Webb, 1961; Coughlin, 1975). Flow through the Batesville Formation is diffuse (Orndorff et al., 1997). The Batesville Formation unconformably overlies the Boone Formation. At some locations, dissolution of the pure limestone of the underlying Boone Formation has allowed the overlying sandstone to collapse, creating sinkholes and other obvious karst features. The overall chemical resistance to weathering of the Batesville Formation preserves these features at the surface and is one of the few settings where karst topography is obvious (Webb, 1961).

The Boone Formation consists primarily of interbedded dark-gray, compact to coarsely crystalline, fossiliferous limestone and chert (Webb, 1961). The limestone portion of the Boone Formation varies from light-gray, coarse-crystalline, crinoidal limestone to dense, dark-gray, non-crystalline limestone (Purdue & Miser, 1916). Dissolution of the Boone Formation occurs along bedding planes, fractures, and joints. The extensive dissolution has created a conduit flow system (Peterson, 1998). The Boone Formation conformably overlies the St. Joe Formation, with the contact defined as the first occurrence of persistent chert (Purdue & Miser, 1916; Van den Heuvel, 1979). Together, the Boone and St. Joe Formation comprise the Springfield Aquifer.

The St. Joe Formation is generally a finely crystalline limestone; it is rich in crinoids and is generally light gray to reddish-brown (Van den Heuvel, 1979; Shanks, 1976; Webb, 1961). The St. Joe Formation is a highly karstic formation that is characterized by a conduit flow system with rapid ground-water transport (Peterson, 1998).

2.1 *Water Quality*

In the present environment of the springs studied, NO_3^- is believed to be very mobile and relatively conservative because of its solubility and ionic form. One potential impeding factor to the nitrate movement would be a decline in redox potential resulting in denitrification. Denitrification refers to the reduction of NO_3^- to N_2O or N_2 through microbial respiratory processes (Canter, 1997; Hallberg & Keeney, 1993). Denitrification generally does not occur when oxic conditions prevail (Canter, 1997). Considering the very open conduit flow, waters of these springs tend to be aerated and should remain aerobic. Therefore, denitrification should not be significant.

Another form of NO_3^- reduction is through chemical denitrification by oxidation of ferrous iron (Hallberg & Keeney, 1993). The low concentrations of iron (Fe) observed in the spring waters suggest that the rate of chemical denitrifcation is not significant as the majority of concentrations were below the detection limit of 0.003 mg/L and the maximum concentrations were less than 0.035 mg/L.

2.2 *Nitrate, Stage, and Specific Conductance Link*

The use of stage as a proxy for NO_3-N concentrations is drawn from the concept that as the volume of water increases in a spring, the NO_3-N load capacity of the spring also increases. Although stage does not directly provide the amount of discharge from the springs, stage does provide a connection to volume based upon a given stage vs. discharge relationship for each spring. The stage also provides a method of incorporating seasonal variation. Stage tends to be lower in the dry periods of the summer and the winter; whereas, in the spring and late-fall, stage tends to be higher.

Use of SP as a proxy for NO_3-N is based on two concepts. First, Adamski & Steele (1988) found a positive correlation between the concentration of calcium and the concentration of NO_3-N within springs of northwest Arkansas. This is important because in the spring water of northwest Arkansas, as with any water discharging from limestone, calcium has primary control on SP. Calcium plays a large role in the SP because of the contribution calcium makes to the ionic strength of the spring water. Ionic strength is the measure of the total concentration of charged species in solution (Fetter, 1994),

$$I = \frac{1}{2}\sum_{i=1}^{n} m_i z_i^2 \tag{1}$$

where I is the ionic strength, m_i is the molality of the ith ion, and z_i is the charge of the ith ion. The high concentration of calcium in springs of northwest Arkansas and the charge of 2^+ for this ion results in water whose ionic strength is essentially controlled by calcium. Using the empirical relationship

$$I = (1.88 * 10^{-5})*SP \tag{2}$$

where I is the ionic strength and SP is the specific conductance, the relationship between ionic strength and SP can be illustrated (Jacobson & Langmuir, 1970). When the ionic strength is dominated by calcium, the SP is also dictated by the concentration of calcium. Hence, if the concentrations of calcium and NO_3-N have a positive correlation, then SP and NO_3-N should also show a correlation.

The second concept also relies on the ionic strength of the water. NO_3-N is included in the calculation of ionic strength. But with the low concentration of NO_3-N in the spring waters and with an ionic charge of 1^-, NO_3-N is not the dominant ion influencing ionic strength in the waters. However, a direct relationship between NO_3-N and SP is established.

3 METHODOLOGY

3.1 *Water Quality*

Each spring was equipped with a Sigma Data Logger model 950 and a tipping-bucket rain gauge. The logger continuously recorded stage, temperature, rainfall, and SP on 15-minute intervals. Data were downloaded monthly. Stafford Spring was instrumented in the winter of 1996. The specific conductance probes were also installed in the winter of 1996 at all sites; all other equipment was installed in 1995.

Storm events were sampled using a Sigma Streamline 800 SL automatic sampler in December 1995, April 1996, and February, May, June, and August 1997. The automatic sampler collected two-liter samples at discrete time intervals throughout the storm events. Samples were retrieved every 24 hours for 2 days, and replaced with a sampler base containing clean bottles. During transportation, the samples were chilled with ice. The samples were split and sent to the Arkansas Water Quality Laboratory for NO_3-N analysis. The Arkansas Water Quality Laboratory used either the EPA/300.0 (ion chromatography) or EPA/353.2 (cadmium reduction/colorimetric) method for NO_3-N analysis. The accuracy of both the EPA/300.0 and EPA/353.2 methods is 0.01 mg/L. Specific conductance of each sample was also recorded in the laboratory using a digital-field specific conductance meter with an accuracy of 0.1 microsiemens per centimeter (μS/cm).

3.2 *Nitrate-Nitrogen Extrapolation*

To simulate NO_3-N concentrations in ground water, present models depend on geological, pedological, or agricultural variables such as the thickness of the aquifer, the permeability of the formation and the overlying soil, the land-use patterns, the soil characteristics, the net water balance, the concentration of NO_3-N on the land surface, and the residence time (Canter, 1997; Farajalla et al., 1993; Cruciani, 1987; Buam, 1994). These variables are not readily available or are gross approximations. This investigation uses a different approach. For this analysis, the independent variables are parameters of the spring water that are readily available and are continuously monitored.

The parameters, stage and SP, were used in conjunction with regression analysis to determine whether a relationship exists between the NO_3-N concentrations, stage, and SP of the springs. Three types of regression analysis were used to determine the best relationship. The methods used included multiple regression (M.R.), piecewise or step multiple regression (Step M.R.), and logarithmic regression (L.R.).

Multiple regression was used to determine the overall relationship of NO_3-N with respect to stage and SP. Following Koch & Link (1971), the equation employed for the analysis was:

$$y = \beta_1 x_1 + \beta_2 x_2 + C \tag{3}$$

where x_1 and x_2 were the independent variables representing stage (x_1) and specific conductance (x_2), NO_3-N concentration is the dependent variable (y), and β_1, β_2, and C are the unknowns. The goal of the regression is to minimize the equation:

$$\sum_{1}^{n}\left(y-\left[C+\beta_1 x_1+\beta_2 x_2\right]\right)^2 \tag{4}$$

for n data points.

When Equation (4) is differentiated with respect to C, β_1, and β_2, and each derivative is set equal to zero, the following linear system is formed:

$$nC + \beta_1 \sum_{1}^{n} x_1 + \beta_2 \sum_{1}^{n} x_2 = \sum_{1}^{n} y$$

$$C \sum_{1}^{n} x_1 + \beta_1 \sum_{1}^{n} x_1^2 + \beta_2 \sum_{1}^{n} x_1 x_2 = \sum_{1}^{n} y x_1 \tag{5}$$

$$C \sum_{1}^{n} x_2 + \beta_1 \sum_{1}^{n} x_2 x_1 + \beta_2 \sum_{1}^{n} x_2^2 = \sum_{1}^{n} y x_2$$

This linear system leaves three equations and three unknowns that can be solved through substitution. For this study, the calculations were performed using algorithms incorporated into a spreadsheet.

The piecewise or step multiple regression uses the same basic theory as the multiple regression, except for one major difference. Instead of developing a single equation for the entire set of data, multiple equations are developed based upon differences in the relationship between one of the independent variables and the dependent variable.

The concept can best be explained by viewing the relationship between stage and NO_3-N concentrations for Stafford Spring (Fig. 3). Note that points exist where the slope of the line dramatically changes. The points at which the slope of the line changes becomes a break point, identifying where a new piece or step of the relationship exists. The

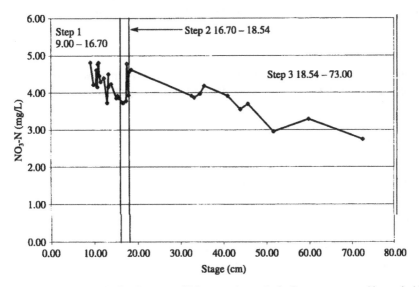

Figure 3. Step determination for step multiple regression analysis. Steps are separated by vertical lines. Data are from Stafford Spring.

sets of data between the break points represent the piece or step. Each piece or step was analyzed as an individual unit using the multiple regression technique previously described. When examining Figure 3, notice that there are three pieces or steps, thus for the numerical analysis of Stafford Spring there were three equations developed to define the relationship between NO_3-N concentrations, stage, and SP. For the other sites, Braly Spring was defined by three equations, Decatur Spring was characterized by five equations, and Little Wildcat Spring and Tanyard Spring each had four equations.

Another possible regression technique used to simulate concentrations is logarithmic regression. Spooner (1996) uses logarithmic regression to quantify a non-linear relationship that can exist with water parameters. Based upon Spooner's (1996) work, the logarithmic equation

$$y = \beta_1 * \beta_2^x \tag{6}$$

with x as the stage, y as the NO_3-N concentration, and β_1 and β_2 as unknowns, was used in the regression analysis. Note that the equation requires only one independent variable, which is the stage. Specific conductance was not involved with the logarithmic regression.

To create a system of linear equations, the natural logarithm of both sides of the above equation is taken creating the equation

$$\ln(y) = \ln(\beta_1) + \ln(\beta_2)x \tag{7}$$

By substituting $\ln(y) = A$, $\ln(\beta_1) = W_{11}$, and $\ln(\beta_2) = W_{21}$, into Equation (7), the resulting basic linear equation is

$$A = W_{11} + W_{21}x \tag{8}$$

This equation can then be solved through linear regression.

Following Spooner's reasoning, two other logarithmic regression techniques where also examined. Mathews (1992) describes the two techniques. The first technique employs the equation

$$y = \beta_1 x \exp(-\beta_2 x) \tag{9}$$

with x as the stage, y as the NO_3-N concentration, and β_1 and β_2 as unknowns. To create a linear system, the natural logarithm is taken of both sides of Equation (9) resulting in equation

$$\ln\left(\frac{y}{x}\right) = \ln(\beta_1) - \beta_2 x \tag{10}$$

By substituting $\ln(y/x) = A$, $\ln(\beta_1) = W_{12}$, and $-\beta_2 = W_{22}$, into Equation (10) the basic linear Equation (11) is defined

$$A = W_{12} + W_{22}x \tag{11}$$

Equation (11) can then be solved through linear regression.

The final technique of Mathews (1992) uses the base equation

$$y = \beta_1 * x^{\beta_2} \tag{12}$$

with x as the stage, y as the NO_3-N concentrations, and β_1 and β_2 as unknowns. To create a linear system, the natural logarithm of both sides of Equation (12) is taken to create Equation (13)

$$\ln(y) = \ln(\beta_1) + \beta_2 \ln(x) \tag{13}$$

By substituting $\ln(y) = A$, $x = \ln(x)$, $\ln(\beta_1) = W_{13}$, and $\beta_2 = W_{23}$, into Equation (13), basic linear equation (14) is defined

$$A = W_{13} + W_{23}x \tag{14}$$

Again, Equation (14) can then be solved through linear regression.

4 RESULTS

4.1 *Water Sampling*

Storm events were sampled in December 1995, April 1996, and February, May, June, and August 1997. Note that not all springs were sampled during each event. All five springs produced data sets that were reliable and complete. The results of the sampling are shown for each spring in Table 1 through Table 5.

4.2 *Simulation of Nitrate Concentrations in Spring Systems*

The regression analysis was performed to simulate NO_3-N concentration at each spring. The models were created using the specific conductance and/or stage data for each spring and M.R., Step M.R., or L.R.. Although only one L.R. equation is listed per spring (Table 6), all three L.R. techniques were employed in the analysis of each spring. The L.R. method that produced the highest correlation coefficient for the spring is listed in Table 6. The observed variations among the three L.R. methods suggest different physical characteristics that affect the stage-NO_3-N relationship of the spring. The modeling equations and statistical results of the analysis are provided in Table 6.

For the Step M.R. analysis, the individual step equations were grouped and viewed as one curve over the entire set of data; therefore, the average error value and the correlation coefficient (r^2) are representative of the entire set of data and not of the individual steps. NO_3-N concentrations simulated using the generated curves are shown in Tables 1-5.

5 DISCUSSION

Although all of the springs are unique in their physical and chemical characteristics (Orndorff et al., 1997), they all show a significant stage and SP response to precipitation events. To illustrate a spring's response to precipitation events, Stafford Spring will be used as an example. Stafford Spring was chosen because of the similarities to the other springs discharging from the Boone Formation. Stafford Spring was also chosen because it illustrates how stage and SP can be used as a proxy for springs that exhibit low varia-

Table 1. Water quality results for Braly Spring.

Sample number	Date	Time	NO$_3$-N (mg/L)	Stage (cm)	Specific conductance (µS/cm)	Step M.R. - NO$_3$-N (mg/L)	M.R. - NO$_3$-N (mg/L)	L.R. - NO$_3$N (mg/L)
12189533	12/17/95	6:05 PM	11.01	3.00	311	10.48	9.93	11.64
12189534	12/17/95	10:35 PM	10.98	2.89	311	10.56	9.91	11.62
12189535	12/17/95	3:35 AM	10.97	3.21	312	10.33	10.02	11.69
12189536	12/17/95	7:35 AM	10.94	3.32	311	10.24	9.99	11.71
12189537	12/18/95	2:45 PM	10.68	3.21	316	10.37	10.22	11.69
12189538	12/18/95	6:15 PM	10.66	3.63	321	10.10	10.55	11.79
12189539	12/19/95	12:15 PM	10.19	4.37	317	9.50	10.49	11.96
12189540	12/19/95	1:15 PM	10.56	4.16	314	9.63	10.30	11.91
961390	04/22/96	2:30 AM	10.41	5.63	313	8.50	10.53	12.26
961393	04/22/96	12:30 PM	12.13	9.21	326	13.53	11.86	13.17
961395	04/23/96	2:00 AM	14.38	13.01	340	15.74	13.28	14.19
961433	04/23/96	8:35 AM	15.57	13.64	349	16.44	13.85	14.37
961435	04/23/96	9:05 PM	16.68	14.37	358	17.19	14.44	14.58
961436	04/24/96	7:05 AM	16.29	12.89	354	16.40	13.96	14.16
971584	02/20/97	10:30 PM	12.39	1.88	293	11.15	8.82	11.39
971586	02/21/97	12:45 PM	12.18	1.88	321	11.43	10.22	11.39
971629	02/21/97	7:00 PM	11.98	2.59	303	10.71	9.45	11.55
971630	02/22/97	1:00 AM	10.16	8.14	285	11.05	9.61	12.89
971631	02/22/97	7:00 AM	10.43	7.67	290	11.11	9.77	12.77
971632	02/22/97	1:00 PM	12.70	9.44	311	12.87	11.15	13.22
971633	02/22/97	6:00 PM	13.79	10.97	324	14.13	12.10	13.63
971682	02/23/97	4:30 PM	13.25	9.32	309	12.72	11.03	13.19
972452	06/12/97	9:33 PM	12.77	6.82	317	12.98	10.96	12.56
972453	06/13/97	5:33 AM	13.01	6.50	319	13.09	10.99	12.48
972454	06/13/97	10:52 AM	13.04	6.90	314	12.90	10.82	12.58
972455	06/13/97	11:52 AM	12.80	6.98	315	12.90	10.89	12.60
972456	06/13/97	1:22 PM	12.67	7.06	315	12.89	10.90	12.62
972457	06/13/97	9:52 PM	12.86	6.90	315	12.92	10.87	12.58
972458	06/14/97	12:52 PM	12.50	6.98	315	12.90	10.89	12.60
972479	06/14/97	7:12 PM	12.77	6.66	315	12.97	10.82	12.52
972480	06/14/97	11:12 PM	12.72	6.42	316	13.04	10.83	12.46
972481	06/15/97	6:12 AM	12.67	6.55	316	13.01	10.86	12.49
972482	06/15/97	9:12 PM	12.54	6.58	316	13.01	10.86	12.50
972483	06/16/97	9:56 AM	12.66	6.50	318	13.07	10.94	12.48
972504	06/17/97	8:40 AM	12.49	7.38	324	13.00	11.41	12.70
972520	06/18/97	4:03 PM	13.23	6.82	333	13.30	11.76	12.56
980445	08/17/97	7:15 AM	12.13	2.33	323	11.11	10.40	11.49
980475	08/18/97	2:25 PM	12.33	2.25	333	11.27	10.89	11.47
980476	08/18/97	5:22 PM	12.19	2.33	332	11.20	10.85	11.49
980477	08/19/97	1:52 AM	12.34	2.41	332	11.14	10.87	11.51
980478	08/19/97	5:52 AM	12.42	2.49	333	11.09	10.93	11.52
980487	08/19/97	5:42 PM	11.98	2.57	328	10.98	10.70	11.54
980488	08/19/97	8:12 PM	12.40	2.49	318	10.94	10.18	11.52
980489	08/19/97	9:42 PM	12.38	2.49	318	10.94	10.18	11.52
980490	08/20/97	1:42 AM	12.28	2.57	318	10.88	10.20	11.54
980491	08/20/97	4:42 AM	12.33	2.33	318	11.06	10.15	11.49
980492	08/20/97	2:15 PM	12.34	2.33	321	11.09	10.30	11.49

Table 2. Water quality results for Decatur Spring.

Sample number	Date	Time	NO$_3$-N (mg/L)	Stage (cm)	Specific conductance (μS/cm)	Step M.R. - NO$_3$-N (mg/L)	M.R. - NO$_3$-N (mg/L)	L.R. - NO$_3$N (mg/L)
961412	04/22/96	2:00 PM	2.83	12.50	264	2.85	2.61	3.86
961413	04/22/96	7:30 PM	2.98	12.87	262	2.89	2.70	3.86
961414	04/23/96	7:30 AM	3.89	14.34	257	3.86	2.98	3.82
961415	04/23/96	1:30 PM	3.45	18.46	244	3.26	3.70	3.54
961442	04/23/96	2:47 PM	3.32	18.64	243	3.31	3.75	3.52
961443	04/23/96	7:17 PM	3.04	18.01	240	3.22	3.84	3.58
961444	04/23/96	12:17 AM	2.95	17.45	238	3.13	3.89	3.63
961445	04/24/96	9:17 AM	2.86	15.98	243	2.74	3.62	3.74
971588	02/19/97	7:15 PM	3.63	9.71	245	3.66	3.23	3.75
971589	02/19/97	10:45 PM	3.63	10.42	244	3.64	3.30	3.80
971590	02/20/97	12:45 AM	3.61	9.90	246	3.62	3.19	3.77
971591	02/20/97	4:45 AM	3.67	10.51	246	3.58	3.23	3.81
971592	02/20/97	9:45 AM	3.77	13.65	247	3.76	3.34	3.85
971593	02/20/97	2:45 PM	4.19	14.69	250	4.21	3.27	3.81
971624	02/20/97	9:05 PM	3.92	19.05	211	3.87	5.05	3.49
971625	02/21/97	3:05 AM	5.06	10.25	202	5.07	4.97	3.79
971626	02/21/97	9:05 AM	6.75	9.46	214	6.73	4.45	3.73
971627	02/22/97	3:05 PM	7.07	9.55	218	7.12	4.30	3.74
971628	02/22/97	4:22 PM	7.11	9.46	221	7.08	4.17	3.73
971683	02/25/97	6:00 PM	5.63	8.76	223	5.64	4.06	3.65
972447	06/12/97	2:04 AM	2.40	6.20	234	2.52	3.49	3.17
972448	06/12/97	7:04 AM	2.62	6.54	234	2.51	3.51	3.25
972449	06/12/97	1:03 PM	2.47	6.67	237	2.54	3.39	3.29
972450	06/12/97	8:03 PM	2.42	5.86	235	2.54	3.43	3.08
972451	06/13/97	4:04 AM	2.67	6.81	234	2.50	3.52	3.32
972463	06/13/97	9:16 AM	2.48	6.54	234	2.51	3.51	3.25
972464	06/13/97	10:16 AM	2.45	8.03	233	2.44	3.62	3.55
972465	06/13/97	11:46 AM	2.46	6.54	233	2.49	3.55	3.25
972466	6/13/97	1:16 PM	2.47	6.54	234	2.51	3.51	3.25
972467	6/13/97	5:16 PM	2.45	7.69	233	2.45	3.60	3.49
972468	6/14/97	5:16 AM	2.46	6.13	234	2.52	3.49	3.15
972469	6/14/97	4:48 PM	2.44	6.40	233	2.50	3.54	3.22
972505	6/17/97	10:03 AM	2.58	6.47	246	2.67	3.02	3.24
972521	6/18/97	2:44 PM	2.72	6.54	247	2.68	2.99	3.25
980363	8/11/97	1:53 PM	2.94	4.50	245	2.73	2.97	2.64
980364	8/11/97	3:23 PM	2.68	5.59	242	2.65	3.14	3.00
980365	8/11/97	4:53 PM	2.71	6.60	244	2.64	3.11	3.27
983066	8/11/97	6:23 PM	2.68	6.67	248	2.69	2.95	3.29
980367	8/11/97	8:23 PM	2.76	6.54	250	2.72	2.87	3.25
980368	8/12/97	5:23 AM	3.02	5.52	261	2.91	2.38	2.98
980406	8/12/97	12:03 PM	2.96	3.42	257	2.93	2.43	2.19
980407	8/12/97	2:06 PM	2.92	4.17	257	2.90	2.47	2.51
980408	8/12/97	3:36 PM	2.85	4.64	256	2.87	2.53	2.69
980409	8/12/97	9:06 PM	2.84	4.37	256	2.88	2.52	2.59
980410	8/13/97	4:06 AM	2.87	3.63	258	2.94	2.40	2.28
980411	8/13/97	9:06 AM	2.85	3.42	262	3.00	2.23	2.19
980432	8/14/97	2:55 PM	2.83	3.02	242	2.74	3.01	1.99

Table 3. Water quality results for Little Wildcat Spring.

Sample number	Date	Time	NO$_3$-N (mg/L)	Stage (cm)	Specific conductance (μS/cm)	Step M.R. - NO$_3$-N (mg/L)	M.R. - NO$_3$-N (mg/L)	L.R. - NO$_3$N (mg/L)
950671	12/17/95	3:15 PM	4.83	16.14	263	6.26	3.84	4.71
950672	12/17/95	6:15 PM	5.34	16.14	287	6.50	4.08	4.71
950673	12/18/95	12:45 AM	4.79	17.32	250	3.63	4.11	4.93
960674	12/18/95	4:45 AM	4.73	17.55	247	3.72	4.16	4.98
960675	12/18/95	2:17 PM	4.83	16.73	262	5.03	4.03	4.82
960676	12/18/95	11:47 PM	5.09	16.37	272	5.86	4.01	4.75
960677	12/19/95	8:47 AM	4.91	16.14	243	6.06	3.64	4.71
960678	12/19/95	2:47 PM	4.85	16.14	237	6.00	3.58	4.71
961389	4/22/96	4:30 AM	4.79	19.44	261	4.09	4.94	5.36
961392	4/22/96	7:30 AM	5.78	20.03	259	4.27	5.12	5.48
961394	4/22/96	2:00 PM	5.12	19.91	219	4.64	4.68	5.46
961396	4/23/96	5:00 AM	6.11	20.73	195	5.10	4.72	5.64
961437	4/23/96	9:23 AM	6.06	20.26	194	4.98	4.55	5.53
961438	4/23/96	1:23 PM	5.74	20.14	193	4.96	4.50	5.51
961439	4/23/96	6:23 PM	5.99	20.14	193	4.96	4.50	5.51
961440	4/24/96	3:23 AM	6.35	19.91	198	4.85	4.47	5.46
971542	2/20/97	3:08 AM	5.67	16.13	231	5.95	3.51	4.71
971541	2/20/97	7:07 AM	5.67	16.00	228	6.19	3.44	4.68
971540	2/20/97	10:50 AM	6.10	16.23	237	5.80	3.61	4.73
971543	2/20/97	1:04 PM	5.66	16.23	237	5.80	3.61	4.73
971657	2/21/97	2:34 PM	5.04	19.25	188	4.77	4.14	5.32
971658	2/21/97	7:57 PM	5.23	18.90	187	4.68	4.02	5.25
971659	2/21/97	11:20 PM	5.28	19.01	189	4.69	4.07	5.27
971684	2/25/97	10:50 AM	6.04	17.74	221	4.03	3.96	5.01
972443	6/12/97	9:58 AM	5.44	14.03	292	6.46	3.41	4.33
972444	6/12/97	4:58 PM	5.49	14.15	297	6.79	3.50	4.35
972445	6/13/97	12:58 AM	5.56	14.27	300	7.00	3.57	4.37
972446	6/13/97	7:16 AM	5.40	14.15	292	6.49	3.45	4.35
972459	6/13/97	8:08 AM	5.39	14.15	286	6.13	3.39	4.35
972461	6/13/97	9:38 PM	5.39	14.27	288	6.28	3.45	4.37
972462	6/14/97	2:38 AM	5.46	14.27	287	6.22	3.44	4.37
972484	6/14/97	8:46 PM	5.40	14.38	289	6.37	3.50	4.39
972485	6/15/97	5:46 AM	5.43	14.38	291	6.49	3.52	4.39
972486	6/15/97	11:46 PM	5.34	14.73	288	6.40	3.61	4.45
972487	6/16/97	11:27 AM	5.24	14.73	284	6.16	3.57	4.45
972506	6/17/97	11:30 AM	4.62	16.70	246	4.93	3.86	4.81
972522	6/18/97	8:30 AM	4.54	16.81	234	4.58	3.78	4.83
972530	6/19/97	1:55 PM	4.86	16.46	243	5.38	3.75	4.77
980446	8/17/97	8:15 AM	2.01	14.48	232	2.97	2.96	4.41
980447	8/17/97	9:45 AM	2.09	16.26	231	5.69	3.56	4.73
980448	8/17/97	11:15 AM	2.48	16.64	250	5.10	3.88	4.80
980449	8/17/97	12:45 PM	3.24	16.76	274	5.08	4.16	4.82
980450	8/17/97	4:45 PM	3.60	16.64	259	5.19	3.97	4.80
980451	8/18/97	4:45 AM	3.00	16.38	234	5.46	3.63	4.75
980468	8/18/97	1:56 PM	2.94	15.62	241	2.50	3.44	4.61
980469	8/18/97	4:26 PM	3.12	15.49	245	2.62	3.44	4.59
980470	8/18/97	7:56 PM	3.28	15.37	250	2.81	3.44	4.57
980471	8/18/97	9:56 PM	3.19	15.37	254	3.13	3.48	4.57
980472	8/19/97	4:56 AM	3.48	15.11	263	3.43	3.49	4.52
980473	8/19/97	3:56 PM	3.99	14.86	270	3.57	3.47	4.48

Table 4. Water quality results for Stafford Spring.

Sample number	Date	Time	NO_3-N (mg/L)	Stage (cm)	Specific conductance (μS/cm)	Step M.R. - NO_3-N (mg/L)	M.R. - NO_3-N (mg/L)	L.R. - NO_3N (mg/L)
971534	2/20/97	4:43 AM	4.51	13.25	390	4.19	4.28	4.30
971533	2/20/97	9:13 AM	2.95	51.63	260	3.32	3.35	3.36
971532	2/20/97	2:43 PM	3.97	34.61	288	3.91	3.82	3.75
971531	2/20/97	4:15 PM	3.87	33.26	280	4.01	3.88	3.78
971609	2/20/97	6:00 PM	3.92	32.67	281	4.03	3.90	3.80
971610	2/20/97	9:00 PM	3.55	44.02	259	3.65	3.60	3.53
971611	2/20/97	11:00 PM	2.74	72.59	225	2.59	2.77	2.94
971612	2/21/97	2:00 AM	3.29	59.85	231	3.11	3.15	3.19
971613	2/21/97	5:30 AM	3.69	45.81	229	3.72	3.60	3.49
971614	2/21/97	10:00 AM	3.91	41.03	236	3.89	3.73	3.60
971615	2/21/97	3:30 PM	4.18	35.56	245	4.08	3.88	3.73
972085	5/2/97	11:50 AM	4.16	13.25	406	4.30	4.25	4.30
972086	5/2/97	1:50 PM	4.23	13.85	405	4.23	4.23	4.28
972087	5/2/97	3:50 PM	4.62	18.54	329	4.89	4.24	4.16
972088	5/2/97	6:50 PM	4.79	17.53	348	4.36	4.24	4.18
972089	5/2/97	10:50 PM	4.56	18.03	361	4.25	4.19	4.17
972090	5/3/97	9:50 AM	3.93	17.93	373	4.03	4.17	4.17
972105	5/3/97	4:20 PM	3.78	17.33	365	3.58	4.21	4.19
972106	5/3/97	7:50 PM	3.73	16.70	368	3.68	4.22	4.21
972107	5/3/97	9:50 PM	3.73	16.51	370	3.70	4.24	4.21
972108	5/4/97	4:50 AM	3.85	15.74	373	3.80	4.23	4.23
972109	5/4/97	9:50 AM	3.91	15.34	380	3.90	4.24	4.24
972110	5/4/97	3:03 PM	3.85	15.04	378	3.91	4.25	4.25
972123	5/7/97	8:46 AM	3.72	12.95	404	4.32	4.26	4.31
980376	8/12/97	3:46 PM	4.22	10.40	351	4.22	4.45	4.38
980377	8/12/97	7:16 PM	4.21	9.85	347	4.25	4.48	4.39
980378	8/13/97	1:46 AM	4.15	10.73	362	4.26	4.42	4.37
980379	8/13/97	6:16 AM	4.29	11.51	380	4.31	4.36	4.35
980380	8/13/97	11:46 AM	4.39	12.23	383	4.24	4.33	4.33
980381	8/13/97	1:41 PM	4.46	11.18	393	4.43	4.34	4.36
980417	8/13/97	3:30 PM	4.52	10.84	401	4.52	4.34	4.37
980418	8/13/97	5:00 PM	4.55	10.73	409	4.59	4.32	4.37
980419	8/13/97	6:30 PM	4.61	10.51	408	4.61	4.33	4.38
980420	8/13/97	11:30 PM	4.77	10.84	415	4.62	4.31	4.37
980421	8/14/97	7:30 AM	4.81	11.01	418	4.62	4.30	4.36
980422	8/14/97	12:30 PM	4.82	9.13	417	4.82	4.36	4.41

tions in the NO_3-N concentrations. Although NO_3-N variability is low at Stafford Spring compared with Tanyard Spring and Braly Spring, the regression analysis and subsequent equations are effective in simulating NO_3-N concentrations.

The low variability in NO_3-N concentrations observed at Stafford, Spring, Decatur Spring, and Little Wildcat Spring strengthen the use of stage and SP as a proxy for NO_3-N. The measured NO_3-N concentrations have an error of roughly 0.2% which is much lower than the error (2.0 to 15.0%) associated with the Step M.R. method. Additionally, the total range of NO_3-N concentrations at each spring is larger than the error associated with the Step M.R. method. Therefore, NO_3-N concentration can be simulated by using stage and SP data and can serve as a proxy for springs with little variation of NO_3-N con-

Table 5. Water quality results for Tanyard Spring.

Sample number	Date	Time	NO₃-N (mg/L)	Stage (cm)	Specific conductance (µS/cm)	Step M.R. - NO₃-N (mg/L)	M.R. - NO₃-N (mg/L)	L.R. - NO₃N (mg/L)
960659	12/17/95	4:00 PM	3.03	20.69	199	3.08	3.02	3.31
960660	12/17/95	8:45 PM	2.79	20.69	197	2.95	2.53	3.31
960661	12/18/95	1:45 AM	2.66	20.69	197	2.95	2.53	3.31
960662	12/18/95	6:45 AM	2.60	20.69	197	2.95	2.53	3.31
960663	12/18/95	1:45 PM	2.45	20.69	197	2.95	2.53	3.31
960664	12/18/95	5:45 PM	2.56	20.69	180	1.82	-1.59	3.31
960665	12/19/95	12:45 AM	8.14	21.14	222	6.95	8.79	3.53
960666	12/19/95	11:45 AM	6.59	21.25	214	7.00	6.90	3.59
961416	4/22/96	10:00 AM	15.25	22.38	253	15.43	16.84	4.19
961417	4/22/96	3:00 PM	14.84	23.06	238	14.24	13.48	4.58
961418	4/23/96	5:00 AM	15.33	23.73	231	12.93	12.07	4.99
961419	4/23/96	10:00 AM	15.27	23.73	236	14.16	13.28	4.99
961446	4/23/96	12:03 PM	14.83	23.88	236	15.23	13.34	5.08
961447	4/23/96	6:33 PM	13.57	23.95	232	14.80	12.40	5.13
961448	4/24/96	3:33 AM	11.66	23.72	225	11.39	10.61	4.98
961449	4/24/96	8:33 AM	10.90	23.37	225	10.98	10.46	4.76
971538	2/20/97	1:09 AM	4.14	23.26	205	5.34	5.56	4.70
971537	2/20/97	5:09 AM	3.99	24.03	181	2.78	0.06	5.18
971536	2/20/97	10:09 AM	7.85	23.77	219	10.28	9.17	5.01
971539	2/20/97	3:15 PM	8.90	25.97	222	9.01	10.82	6.53
971616	2/20/97	5:02 PM	8.85	25.85	220	8.57	10.28	6.43
971617	2/20/97	6:32 PM	8.67	25.59	217	7.85	9.45	6.24
971618	2/20/97	8:32 PM	8.49	24.68	209	5.68	7.13	5.61
971619	2/20/97	10:32 PM	8.57	27.66	194	6.83	4.74	7.87
971620	2/21/97	1:32 AM	10.26	27.53	206	8.43	7.59	7.76
971623	2/21/97	3:04 AM	10.96	28.56	222	11.90	11.91	8.66
971621	2/21/97	5:32 AM	10.84	27.27	215	9.44	9.67	7.54
971622	2/21/97	10:32 AM	10.96	28.56	217	11.18	10.70	8.66
972092	5/2/97	1:45 PM	3.90	24.68	198	4.08	4.46	5.61
972093	5/2/97	3:45 PM	3.85	24.55	198	3.94	4.40	5.52
972094	5/2/97	6:45 PM	3.85	25.33	198	4.81	4.73	6.05
972095	5/2/97	10:15 PM	3.83	25.59	198	5.10	4.84	6.24
972096	5/3/97	2:45 AM	3.79	25.59	197	4.95	4.59	6.24
972097	5/3/97	8:15 AM	3.74	25.59	197	4.95	4.59	6.24
972099	5/3/97	3:35 PM	3.81	25.33	199	4.95	4.97	6.05
972100	5/3/97	6:35 PM	3.81	25.51	199	5.16	5.05	6.18
972101	5/3/97	8:35 PM	3.82	25.59	198	5.10	4.84	6.24
972102	5/4/97	7:35 AM	3.79	25.65	199	5.31	5.11	6.29
972103	5/4/97	1:05 PM	3.82	24.81	199	4.37	4.75	5.69
972104	5/4/97	2:25 PM	3.79	24.29	199	3.79	4.53	5.35
980371	8/12/97	10:24 PM	3.02	23.30	195	2.62	3.15	4.72
980372	8/13/97	2:24 AM	3.01	23.41	194	2.47	2.95	4.79
980373	8/13/97	7:24 AM	3.02	23.53	194	2.60	3.00	4.86
980374	8/13/97	11:32 AM	2.98	23.06	197	2.92	3.53	4.58
980424	8/13/97	3:45 PM	2.99	22.58	201	3.52	4.30	4.30
980425	8/13/97	4:45 PM	3.08	22.69	200	3.36	4.10	4.36
980426	8/13/97	8:15 PM	3.01	22.70	197	2.54	3.38	4.37
980427	8/13/97	10:15 PM	3.09	23.06	197	2.92	3.53	4.58
980428	8/14/97	1:15 AM	2.92	22.46	197	2.29	3.28	4.24
980429	8/14/97	5:15 AM	3.03	22.58	203	4.08	4.79	4.30
980430	8/14/97	10:15 AM	3.26	23.65	198	3.83	4.02	4.94
980431	8/14/97	1:15 PM	3.09	22.57	198	2.68	3.57	4.29

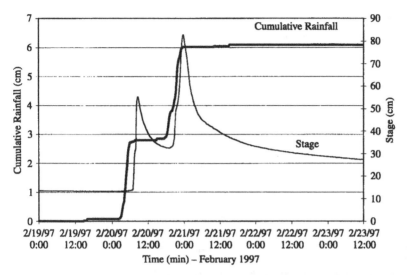

Figure 4. Cumulative rainfall and stage for Stafford Spring following the February 1997 precipitation event.

centrations without being biased by the low variation in NO_3-N concentrations. The accurate simulation of NO_3-N concentrations, where the variation in the concentrations is low, further strengthens the overall relationship among stage, SP, and NO_3-N.

A spring's response following a precipitation event can provide information about the flow system. For Stafford Spring, the increase in the stage and the lag time between the precipitation event and the stage change is shown in Figure 4. The short lag time for Stafford Spring is typical of the springs discharging from the Boone Formation and is consistent with conduit flow. Springs discharging from the Batesville Sandstone and Pitkin Formation have a longer lag time, which is more consistent with diffuse flow. Using stage for comparison, the response of NO_3-N and SP following a precipitation event can also be examined (Fig. 5a). The overall relationship exhibited for Stafford Spring is unique to Stafford Spring, but similar trends are noted in the other springs (Fig. 5b).

The three regression analysis techniques produced different equations, error values, and correlation coefficients for each spring (Table 6). As a complement to Table 6, graphical representations of the measured data for Stafford Spring versus the M.R., Step M.R., and L.R. simulated NO_3-N data are provided in Figures 6a-6c.

A closer comparison of the graphs in Figures 6a and 6c shows how the M.R. and the L.R. both are adequate simulators for NO_3-N concentrations, but they do not accurately predict the extreme high and low concentrations. However, the Step. M.R. (Fig. 6b) is both more accurate and precise in simulating NO_3-N concentrations, especially with the simulation of the outlying NO_3-N concentrations. Both the set of graphs and the tabular data demonstrate that the Step M.R. provides the best approximation for the NO_3-N concentrations at Stafford Spring, as well as at the other springs. Since the resulting Step M.R. equations generate the most accurate and precise simulated NO_3-N concentrations, and because the Step M.R. curves provide the best correlation (among the three regression analysis methods) and lowest error for each spring, the primary focus of the remaining discussion will be use of the Step M.R. method.

Table 6. Regression analysis results.

Spring	Regression method	Equation	Percent error (%)	r^2
Braly	M.R.	$y = 0.05(S.P.) + 0.19(Stage) - 6.19$	4.94	0.70
	Step M.R. (0-5.631 cm)	$y = 0.01(S.P.) - 0.76(Stage) + 9.65$	2.49	0.92
	Step M.R. (5.631-7.379 cm)	$y = 0.02(S.P.) - 0.21 (Stage) + 8.07$		
	Step M.R. (7.379 cm +)	$y = 0.05(S.P.) + 0.40(Stage) - 6.46$		
	L.R.	$y = 1.02^{Stage} * 10.97$	6.48	0.47
Decatur	M.R.	$y = -0.04(S.P.) + 0.05(Stage) + 12.54$	22.98	0.29
	Step M.R. (0-8.026 cm)	$y = 0.01(S.P.) - 0.03(Stage) - 0.47$	2.20	0.99
	Step M.R. (8.026-9.547 cm)	$y = 0.05(S.P.) + 2.21 (Stage) - 24.85$		
	Step M.R. (9.547-12.502 cm)	$y = -0.03(S.P.) - 0.07(Stage) + 12.50$		
	Step M.R. (12.502-14.692 cm)	$y = -0.02(S.P.) + 0.49 (Stage) + 2.89$		
	Step M.R. (14.692 cm +)	$y = -0.01(S.P.) + 0.22(Stage) + 2.85$		
	L.R.	$y = 0.84 * Stage * exp(-0.08 * Stage)$	20.44	0.16
Little Wildcat	M.R.	$y = 0.01(S.P.) + 0.34(Stage) - 4.28$	22.00	0.14
	Step M.R. (0-14.727 cm)	$y = 0.06(S.P.) + 0.26(Stage) - 14.71$	11.04	0.66
	Step M.R. (14.727-15.621 cm)	$y = 0.08(S.P.) + 1.65(Stage) - 42.55$		
	Step M.R. (15.621-16.812 cm)	$y = 0.01(S.P.) - 2.06 (Stage) + 36.87$		
	Step M.R. (16.812 cm +)	$y = -0.01(S.P.) + 0.27(Stage) + 1.45$		
	L.R.	$y = 1.04^{Stage} * 2.50$	22.65	0.12
Stafford	M.R.	$y = -0.02(S.P) - 0.03(Stage) + 5.49$	6.39	0.60
	Step M.R. (0-16.703 cm)	$y = 0.01(S.P.) - 0.11(Stage) + 2.91$	3.21	0.86
	Step M.R. (16.703-18.542 cm)	$y = -0.02(S.P.) + 0.21(Stage) + 6.43$		
	Step M.R. (18.542 cm +)	$y = -0.01(S.P.) - 0.04(Stage) + 6.79$		
	L.R.	$y = 0.99^{Stage} * 4.68$	6.80	0.61
Tanyard	M.R.	$y = -0.24(S.P) + 0.42 (Stage) - 53.98$	21.43	0.88
	Step M.R. (0-22.380 cm)	$y = 0.07(S.P.) + 5.18(Stage) - 117.25$	15.52	0.94
	Step M.R. (22.380-23.652 cm)	$y = 0.28(S.P.) + 1.06(Stage) - 75.85$		
	Step M.R. (23.652-24.033 cm)	$y = -0.25(S.P.) + 7.36(Stage) - 218.88$		
	Step M.R. (24.033 cm +)	$y = 0.14(S.P.) + 1.12 (Stage) - 52.20$		
	L.R.	$y = 0.0004 * (Stage)^{2.9781}$	35.4	0.64

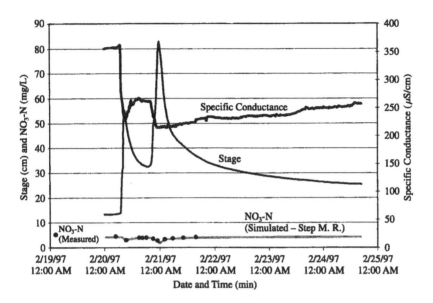

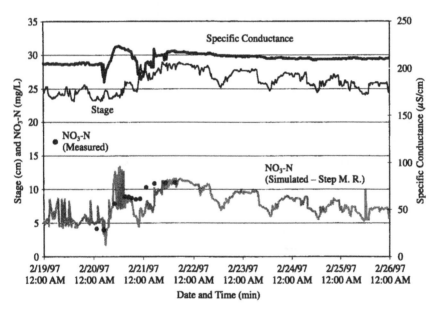

Figure 5. a) Storm hydrograph for Stafford Spring following the February 1997 precipitation event. Hydrograph includes stage, SP, measured NO_3-N concentrations, and simulated Step M.R. NO_3-N concentrations. b) Storm hydrograph for Tanyard Spring following the February 1997 precipitation event. Hydrograph includes stage, SP, measured NO_3-N concentrations, and simulated Step M.R. NO_3-N concentrations.

Based on the correlation coefficients ($r^2 \geq 0.66$), the Step M.R. provides a method for predicting continuous NO_3-N concentrations at a spring. Application of the resulting Step M.R. equations were applied to available data at each spring. Application of continuous stage and SP data to the Step M.R. equations results in a simulated continuous NO_3-N

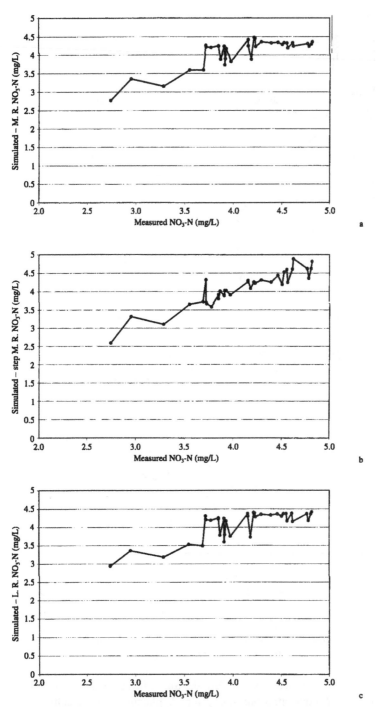

Figure 6. a) Comparison between measured NO$_3$-N concentrations and simulated M.R. NO$_3$-N concentrations. b) Comparison between measured NO$_3$-N concentrations and simulated Step M.R. NO$_3$-N concentrations. c) Comparison between measured NO$_3$-N concentrations and simulated L.R. NO$_3$-N concentration.

curve. Simulated NO_3-N curves were generated at each spring for each storm. Two examples of the curves are presented in Figures 5a and 5b. Figure 5a shows the February 1997 precipitation event at Stafford Spring, and Figure 5b illustrates the February 1997 precipitation event at Tanyard Spring. The calculated error values for the generated curves of all the springs are provided in Table 6, but visual interpretation of the generated curves shows that the Step M.R. curves nearly approximate the true NO_3-N concentrations, further confirming the use of Step M.R. to simulate NO_3-N concentrations. The use of the Step M.R. is not limited to the storm hydrograph. The modeled equations can be applied to the entire set of continuous data to generate simulated NO_3-N concentration records.

Although a goal of the regression analysis was to simulate NO_3-N concentrations, the equations also help to describe hydrogeologic elements of the springs. Examination of Table 6 shows that for the diffuse-flow systems, the three methods of regression analysis produce slightly better results than with the conduit-flow systems. At both the diffuse-flow springs (Tanyard & Braly), the correlation coefficients for the M.R. were 0.70 or better and were higher than the M.R. correlation coefficients for the conduit springs (Decatur, Little Wildcat, & Stafford). The high correlation coefficients for a diffuse-flow system indicate that M.R. may be used as an adequate modeling technique; conversely, M.R. may not be an adequate modeling technique for the conduit-flow systems. ·

An additional piece of information inferred from the equations is which parameter is driving the model prediction of NO_3-N concentrations. The parameter that most significantly influences the approximation can be found using the absolute value ratio of the multipliers for both the SP and the stage. A ratio below 0.1 would indicate that the approximations are based more upon the stage data than the SP data, and a ratio above 0.1 would indicate that SP was more important in the approximation (Pearson, 1974). Normally a ratio of 1 would indicate that both SP and stage are equally significant, but since SP is a magnitude higher, it is necessary to use a ratio of 0.1.

The uses of the ratio can be demonstrated at the five springs. For example, at Decatur Spring the average ratio for the Step M.R. between the multipliers for SP and stage is approximately 0.20 suggesting that SP exerts a greater influence. At Little Wildcat Spring the average ratio for the Step M.R. between the multipliers for SP and stage is approximately 0.09 suggesting slightly greater influence of stage.

After producing the equations for each spring, the equations were interchanged with the other springs to determine if one set of equations is universal. Application of the M.R., Step M.R., and L.R. equations from one spring to the other springs produced correlation coefficient values less than 0.10 and error values consistently higher than 50 percent. The results indicate that the equations are not interchangeable between springs. This implies that different hydrogeologic conditions exist at the individual springs, which then influence the concentration of NO_3-N. Examples of the differences are identified in the equations generated for each spring. The coefficients (multipliers) for SP and stage indicate the type of relationship (direct or inverse) that exists between the parameter and NO_3-N concentration. For example at Stafford Spring, when the stage multiplier is negative, an inverse relationship between stage and NO_3-N concentration exists. Conversely, at Tanyard Spring, the stage multiplier is always positive indicating a direct relationship between stage and NO_3-N concentration. Another example would be at Braly Spring, in which a direct relationship between SP and NO_3-N exists.

The L.R. equations were the poorest at approximating the NO_3-N concentrations at the five sites. Two possible reasons exist to describe the poor performance. The first is that a second variable is needed in order to adequately model the concentration of NO_3-N, and secondly, the NO_3-N concentrations do not have a logarithmic relationship with stage at these sites.

6 CONCLUSIONS

The main purpose of this project was to determine if a relationship existed between the NO_3-N concentrations, stage, and SP. The study focused upon five springs in northwest Arkansas that have been sampled periodically over the past two years. Based on the interpretation of the measured and simulated data, the following conclusions can be made

1. A correlation among NO_3-N, stage, and specific conductance exists at each spring. The resulting relationships among NO_3-N, stage, and specific conductance are unique for each spring. Generally as stage increases and SP decreases, NO_3-N concentrations decrease.

2. The Step M.R. provides the best method for simulation based upon the highest correlation coefficients ($r^2 \geq 0.66$) and lowest error for the prediction of NO_3-N concentrations.

3. The L.R. approach does not appear to be adequate for these data sets. Correlation coefficients are generally less than 0.50 ($r^2 < 0.50$) and the resulting approximations have a high error value.

4. This relationship among SP, stage, and NO_3-N can be used to simulate continuous NO_3-N concentrations at the spring sites.

5. NO_3-N concentrations were extrapolated over the entire set of continuous data.

6. Each spring has a unique set of regression equations.

7. The generated equations provide information regarding the type (direct or inverse) and the strength of the relationship between either stage or SP with the NO_3-N concentrations.

8. The use of Step M.R. is applicable to other karst areas, given background stage, specific conductance, NO_3-N concentration data, and familiarity of a spring's response to storm events.

REFERENCES

Adamski, J.C. & Steele, K.F. 1988. *Agricultural land use effects on ground water quality in the Ozark region, in Agricultural impacts on ground water,* Des Moines. National Water Well Association, Dublin OH, p. 593-614.

Alexander, R.B. & Smith, R.A. 1990. County-level estimates of nitrogen and phosphorus fertilizer use in the United States: U.S. Geological Survey Open-File Report 90-130, 12 p.

Arkansas Agricultural Statistics Service, 1995. Arkansas agricultural statistics for 1995. Arkansas Agricultural Experiment Station, Division of Agriculture, University of Arkansas Report Series 334, Fayetteville, 60 p.

Branner, G.C. 1940. Mineral resources of Benton, Carroll, Madison, and Washington Counties: State of Arkansas, Arkansas Geological Survey, County Mineral Report 2, Little Rock, 56 p.

Buam, E.J. 1994. Predicting groundwater contamination with mathematical models, in Zoller, U. (ed.), *Groundwater contamination and control.* New York: Marcel Dekker, Inc., p. 547-566.

Canter, L.W. 1997. *Nitrates in groundwater:* Boca Raton, FL, CRC Lewis, 263 p.

Coughlin, T.L. 1975. Geologic and environmental factors affecting ground water in the Boone Limestone of north central Washington County, Arkansas (M.S. thesis). Fayetteville, University of Arkansas, 98 p.

Croneis, C. 1930. *Geology of the Arkansas Paleozoic area with especial reference to oil and gas possibilities.* State of Arkansas, Arkansas Geological Survey, Little Rock, 457 p.

Cruciani, C.L.W. 1987. The prediction of nitrate contamination potential using known hydrogeologic properties (M.S. thesis). Milwaukee, University of Wisconsin – Milwaukee, 181 p.

Davis, J.V., Peterson, J.C., Adamski, J.C. & D.A. Freiwald, 1995. Water-Quality assessment of the Ozark Plateaus study unit, Arkansas, Kansas, Missouri, and Oklahoma-Analysis of information on nutrients, suspended sediments, and suspended solids, 1970-93. U.S. Geological Survey Water-Resources Investigations Report 95-4042, 112 p.

Farajalla, N.S., Deyle, R.E., Vieux, B.E. & Canter, L.W. 1993. Correlating nitrate levels in ground water with agricultural land use in Oklahoma, in *Proceedings of 1993 Joint CSCE-ASCE National Conference on Environmental Engineering.* New York, pp. 469-476.

Fetter, C.W. 1994. *Applied hydrogeology,* Third edition. New York: Macmillian College Publishing Company, 691 p.

Goldberg, M.C. 1970. *Agricultural practices and water quality.* Iowa State University Press, Ames, 107 p.

Hallberg, G.R. & Keeney, D.R. 1993. Nitrate, in Alley, W.M. (ed.) *Regional ground-water quality.* New York: Van Nostrand Reinhold, p. 297-322.

Jacobson, R.L. & Langmuir, D. 1970. The chemical history of some spring waters in carbonate rocks. *Ground Water,* 8: 5-9.

Koch, G.S.Jr. & Link, R.E. 1971. *Statistical analysis of geologic data, Volume 2.* New York: Dover Publications, 438 p.

Leidy, V.A. & Morris, E.E. 1990. Hydrogeology and quality of ground water in the Boone Formation and Cotter Dolomite in karst terrain of northwestern Boone County, Arkansas: U.S. Geological Survey Water-Resources Investigation Report 90-4066, 57 p.

Mathews, J.H. 1992. *Numerical methods for mathematics, science, and engineering,* 2nd edition: Prentice Hall, Englewood Cliffs, NJ, 646 p.

Orndorff, H.A., Davis, R.K. & Brahana, J.V. 1997. *Comparison of shallow aquifer flow systems in Northwest Arkansas.* Geological Society of America Abstract with Programs, 29(6): 106 .

Pearson, C.E. 1974. *Handbook of applied mathematics.* New York: Van Nostrand Reinhold Company, 1265 p.

Peterson, E.W. 1998. Movement of nitrate through regolith covered karst, Northwest Arkansas (M.S. thesis). Fayetteville: University of Arkansas, 159 p.

Purdue, A.H. & Miser, H.D. 1916. U.S. Geological Survey Geologic Atlas, Eureka Springs-Harrison Folio, (No. 202), 22 p.

Scott, H.D. 1989. Simulation of the fate of nitrogen from the disposal of poultry litter. University of Arkansas, Fayetteville, Arkansas Water Resources Research Center Publication No. 142, 47 p.

Shanks, J.L. 1976. Petrology of the St. Joe Limestone in its type area, North Central Arkansas (M.S. thesis). Fayetteville, University of Arkansas, 87 p.

Spooner, J. 1996. Water quality monitoring design and statistical analysis for non-point source pollution: *A short course: Arkansas Water Resources Center's Conference on Diversity of Arkansas Water Resource Research,* Fayetteville, Arkansas, April 22 and 23, 1996.

Van den Heuvel, P. 1979. Petrography of the Boone Formation, Northwest Arkansas (M.S. thesis). Fayetteville: University of Arkansas, 75 p.

Webb, J.H. 1961. The Geology of Green Forest Township, Carroll County, Arkansas (M.S. thesis). Fayetteville: University of Arkansas, 59 p.

Surface and ground water mixing, flow paths, and temporal variations in chemical compositions of karst springs

JONATHAN B. MARTIN & SHERYL L. GORDON
University of Florida, Department of Geology, Gainesville, Florida, USA

ABSTRACT: Karst aquifers provide important sources of water, but may be susceptible to contamination from mixing with polluted surface water. Characteristics of mixing, in particular the extent of conduit versus diffuse flow, have been studied through time-series observations of the chemical composition of water discharging from springs (chemographs). Annual and storm chemographs of Ca, Mg, DIC, and alkalinity are constant within error for six closely spaced springs that discharge from the Floridan aquifer in north-central Florida, reflecting equilibration with carbonate rocks of the aquifer. Similar chemographs from other karst aquifers have been interpreted to reflect diffuse flow, but dye trace and cave diving evidence indicate conduit flow is significant in the study area. Chemographs from Floridan aquifer springs may differ from other karst springs because of a shallow water table, which places most conduits in the phreatic zone and increases the volume of ground water available to mix with recharged water. Three of the springs, with previously identified connections to swallets, show temporal variations in Cl and SO_4 chemographs, but little variations in temperature or oxygen chemographs. In contrast, the three other springs show little change in Cl and SO_4 chemographs, but significant and systematic decreases in temperature and oxygen chemographs from summer (rainy season) to winter (dry season). The Cl and SO_4 chemographs reflect mixing with dilute surface water, while the oxygen and temperature chemographs reflect the depth of the source of water. Seasonal NO_3 chemographs vary widely because of contamination. The storm chemographs show minor dilution of Cl and SO_4 concentrations, at the same time as an increase from no detectable NO_3 to ~1.5 mg/kg NO_3, presumably from contamination. These results show that information unique from individual chemographs is provided by simultaneous chemographs of several solutes from groups of karst springs.

1 INTRODUCTION

Karst terraines are estimated to cover 25% of the earth's land surface, and aquifers in karst rocks provide approximately a quarter of the world's population with potable water (Ford & Williams, 1989). Karst aquifers are characterized by heterogeneous permeability, however, complicating standard ground water flow analysis and modeling (e.g. Dreiss, 1983; 1999a; Wicks & Herman, 1994). High vertical permeabilities result from solution features such as sinkholes that allow point recharge of large volumes of surface water. Subsurface flow can be rapid and when coupled with rapid recharge makes karst aquifers susceptible to pollution (e.g. Hess & White, 1988; Felton & Currens, 1994; Iqbal & Krothe, 1995; White & others, 1995; Katz & DeHan, 1996; Pasquarell & Boyer, 1996; Kincaid, 1997; LaMoreaux & others, 1997; Katz & Hornsby, 1998). In addition to problems associated with pollution, mixing of surface and ground waters is a critical control of the natural ground water chemistry (e.g. White & others, 1995; Katz & others, 1995a;b; Greene, 1997; Katz & others, 1977a;b). Surface water is often undersaturated with respect to the soluble karst minerals, and this disequilibrium promotes karstification and engineering problems associated with dissolution (e.g. Upchurch & Lawrence, 1984; Günay & Johnson, 1986; Beck, 1995; LaMoreaux & others, 1997). The important environmental characteristics of karst, coupled with the difficulty of applying standard hydrologic models and techniques, have led to many observational studies of the chemistry of karst waters and its relationship to fluid-solid reactions and mixing.

One technique for identifying the origins of recharged water, and mixing in the subsurface, is through dye trace studies. Such studies are limited, however, by complicated logistics and commonly are restricted to short intervals of time and small ranges of flow conditions (e.g. White, 1988). In order to understand the distribution of flow paths in karst aquifers, however, dye trace studies need to be conducted at a range of flow conditions because of the heterogeneous nature of permeability, conduits, fractures, and intergranular porosity. Injected dyes also can be affected by artifacts caused by dispersion and reactions between the aquifer rocks and the dyes.

The temporal and spatial variations in the natural chemical and isotopic compositions of karst water have also been used to study the mixing of waters (e.g. Pitty, 1968; Shuster & White, 1971; 1972; Ternan, 1972; Hess & White, 1988; Jones & others, 1993; Saunders & Toran, 1994; Sacks & others, 1995; Wicks & others, 1995). For example, mixing of ground and surface water has been measured with the abundances of naturally occurring isotopes of radon, oxygen, and hydrogen (Lakey & Krothe, 1996; Kincaid, 1997). Spatial variations in the chemical composition of karst water have largely been studied with respect to carbonate equilibrium and chemistry (e.g. Wigley & Plummer, 1976; Plummer, 1977; Hunn & Slack, 1983; Chapelle & Knobel, 1986; Chapelle & others, 1988; Sprinkle, 1989; Upchurch, 1992; Katz, 1992). Changes in concentrations of natural components and contaminants, particularly NO_3, have been used to identify how land use practices and downward mixing from the epikarst layer affect the water quality in karst aquifers (Iqbal & Krothe, 1995; Boyer & Pasquarell 1995; 1996). Temporal variations of water hardness, conductivity, and temperature, commonly referred to as chemographs, have been particularly useful for determining the extent of flow within large, high-permeability conduits and small intergranular pore spaces (Pitty, 1968; Shuster & White, 1971; 1972; Hess & White, 1988).

Studies using chemographs generally focus on single, or at most only a few, chemical components (e.g. Shuster & White, 1971; 1972; Hess & White, 1988; Kincaid, 1997). There are few studies of time-series measurements of closely spaced springs from within a single ground water basin. Such studies could provide important information about the heterogeneity of physical and chemical processes over small distances within a single basin. For example, temporal changes in compositions of water discharging from closely spaced springs would reflect mixing and illustrate heterogeneity of flow paths over small distances. Fractionation of the concentrations of various dissolved components, each controlled by different reactions, would identify water-rock reactions. Ultimately, these results could provide the basis for the development of conceptual flow models, and lead to constraints for numerical simulations of physical mixing and chemical reactions within karst aquifers.

This paper focuses on temporal variations in chemical composition of water discharging from six springs along an ~4 km stretch of the Ichetucknee River in north-central Florida (Fig. 1). The springs discharge from the unconfined Floridan aquifer, one of the most productive karst aquifers in the world, and a major water supply for the southeastern U.S. The concentrations of numerous dissolved components (O_2, Cl, SO_4, NO_3, DIC, alkalinity, pH, Mg, and Ca) and the temperature of the water were measured in samples that had been collected approximately every two weeks for one year. Many of these com-

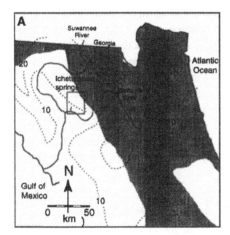

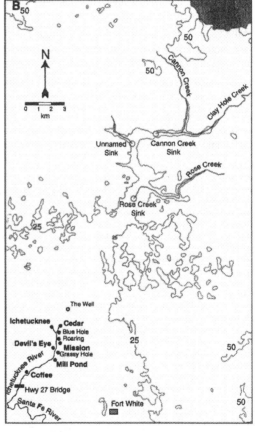

Figure 1. A) Map of north-central Florida showing location of Ichetucknee springs area. Also shown are the two major rivers in the region, the Suwannee and Santa Fe rivers. The shaded region represents the location of confining Hawthorn Group rocks. The dashed lines represent the potentiometric surface of the Floridan aquifer in meters. Modified from Rosenau et al. (1977), Scott (1988), and Sprinkle (1989). B) Topographic map of region surrounding the Ichetucknee springs. Shown are the location of various sinking streams and nine named springs along the Ichetucknee River. The six springs with names in bold were sampled for this study. The 25 m contour approximates the location of the Cody Scarp, and represents the boundary between confined and unconfined Floridan aquifer. Contour interval is 25 m.

ponents were also measured in samples collected at 3 to 12 hour intervals for 372 hours following a major storm. The observed variations in concentrations with time are used to address questions relating to mixing of surface and ground water and the heterogeneity of flow paths within a small region. The temporal variations in chemical composition also reveal the relative importance of recharge from individual storms compared with seasonal variation in recharge.

2 GEOLOGY, CLIMATE, AND HYDROLOGY

2.1 *Regional description*

Rocks are divided into both lithologic and hydrologic units in the Florida carbonate platform because the aquifers cut across formational boundaries (Fig. 2). In general, pre-Miocene rocks are carbonate, with minor siliciclastics, while Miocene and younger rocks are largely siliciclastic with minor and discontinuous carbonate lenses (Scott, 1988; 1992; Groszos & et al., 1992). Miocene-aged siliciclastic rocks make up the Hawthorn Group, which acts as the confining unit for the Floridan aquifer. The Miocene and younger sediments have been removed by erosion from much of the western half of north-central Florida, and as a consequence, the Floridan aquifer is unconfined in this region. The erosional edge of the Hawthorn Group extends across the northeastern corner of the field area, and is a major influence on the geomorphology of the region (Fig. 1A). The north-

System	Series	Hydrostratigraphic Unit	Lithostratigraphic Unit	Lithologic Descriptions
Quaternary	Holocene	Surficial Aquifer System	Undifferentiated Pleistocene-Holocene Sediments	Fine to coarse grained poorly indurated quartz sands with minor amounts or clay. Peat and fresh water carbonates are present in places.
	Pleistocene			
Tertiary	Pliocene	Confining Unit, contains lenses of the Intermediate Aquifer	Miccosukee Fm Cypresshead Fm Nashua Fm	Interbedded clay, silt, sand, and gravel siliciclastics with carbonate content increasing upward.
	Miocene		Hawthorn Group Statenville Fm Coosaqhatchie Fm Markshead Fm Penny Farms Fm St. Marks Fm	Fine to coarse grained quartz sand, silt and clay minerals with occasional phosphate mineral-rich units. The intermediate aquifer is confined to four distinct dolostone units.
	Oligocene	Floridan Aquifer System	Suwannee Limestone	Variable vuggy and muddy limestone.
	Eocene		Ocala Limestone Avon Park Fm Oldsmar Fm	Fossiliferous limestone interbedded with vuggy dolostone. Avon Park acts as a confining unit in places.
	Paleocene	Sub-Floridan Confining Unit	Cedar Keys Fm	Coarsely crystalline, porous dolostone w/minor limestone and evaporites.
Cretaceous and older			Undifferentiated	

Figure 2. Lithostratigraphy and hydrostratigraphy of north-central Florida. Miocene and younger strata, and their hydrostratigraphic units, are confined to the northeastern portion of the Ichetucknee River drainage basin shown in Fig. 1B. Modified from Scott (1988; 1992) and Groszos et al. (1992).

ern region is described as the Northern Highlands because its elevations are typically greater than 50 m above sea level (Puri & Vernon, 1964; Stringfield, 1966; White, 1970; Brooks, 1982). The southwestern region is described as the Gulf Coast Lowlands, with elevations ranging from ~25 m to sealevel. The Gulf Coast Lowlands are dominated by karst plains where Eocene-aged carbonate rocks are exposed. The Northern Highlands are separated from the Gulf Coast Lowlands by a feature called the Cody Scarp, which represents the boundary between the confined and unconfined Floridan aquifer. This boundary is significant because it appears to exert an important control over transmissivities within the aquifer, as well as chemical composition of the ground water (e.g. Upchurch & Lawrence, 1984; Ryder, 1985; Tibbals, 1990; Pucci & others, 1992; Wicks & Herman, 1994).

The potentiometric surface of the Floridan aquifer slopes from the northeast to the southwest, generally following topography (Sprinkle, 1989). There is little surface drainage in the recharge area because the area immediately north of the Ichetucknee River is unconfined (Scott, 1988; 1992; Groszos & others, 1992). The streams in the recharge area flow only during storms, and discharge directly to groundwater through sinkholes (Fig. 1B). The lack of surface drainage indicates that volume of recharge can be estimated simply as precipitation minus evapotranspiration. Therefore, this region is ideally suited to observe how karst aquifers respond to potential recharge at seasonal time scales versus individual storms of relatively short duration.

The climate in the region is humid subtropical. The average annual temperature is ~22°C. On average, the coldest month is January, with average temperatures of 11.4°C and average daily range from 4.8 to 17.8°C. The warmest month is July with average temperatures of 27.1°C and average daily range from 21.3 to 32.8°C. Precipitation averages ~140 cm/yr, and typically ~40% of the precipitation occurs as thunderstorms during June, July, and August. The storms are often locally derived, resulting in large differences in precipitation between closely spaced locations. Additional large storms can occur at times other than the rainy season, either as tropical storms such as hurricanes during the late summer or fall, or as cold fronts during the late fall and winter. Such storms can produce significant rainfall in short amounts of time, and because they occur at normally dry times, they may recharge the aquifer at different rates than summer storms.

2.2 *Hydrology of the Ichetucknee springs*

Over the past decade, several unpublished dye tracing experiments have been performed around the Ichetucknee River for Florida's Department of Environmental Protection. The earliest experiment used rhodamine dye to show a ground water connection between a karst window called The Well and Ichetucknee Spring (Fig. 1). The dye emerged at Ichetucknee Spring in 12 peaks, 15 to 83 hours after injection, indicating flow rates > 2.0 km/day. The rapid flow rate reflects conduit flow, but the long time between the initial and final return of the dye and the separation of the returned dye into several peaks suggest that the karst in this region may be characterized by a mixture of braided conduits, as well as some slow flow through interconnected pores and fractures (these results are discussed in Gordon (1998)). A second tracer experiment injected SF_6, a non-reactive gas with low solubility, into Rose Creek Sink, approximately 10 km north of the Ichetucknee springs (Fig. 1). The study indicated a connection between Rose Creek Sink and all of the springs that were sampled in this study except for Coffee Spring, which was not sampled (Hirth, 1995). The initial return of the SF_6 was not observed during the tracer study, however, suggesting there may have been a problem with contamination from an earlier injection of SF_6 into a water supply well 5 km to the east of the Ichetucknee River.

A third tracer study using fluorescene dye was conducted during the summer of 1997 because of potential problems with the Hirth (1995) study. Fluorescene dye was injected by SCUBA divers directly into a conduit in the Floridan aquifer that leads out of Rose Creek Sink at a depth ~15 m below the surface. Dye emerged first at Mission Spring, and was also detected at Blue Hole, Devil's Eye, Grassy, and Mill Pond springs (Fig. 1). No dye was found at Ichetucknee, Cedar Head, or Coffee springs, or at The Well. Somewhat surprisingly, these two groups of springs are separated geographically, with Mission, Devil's Eye, and Mill Pond springs located in the middle reaches of the river, Ichetuck-nee and Cedar springs geographically closest to Rose Creek Sink at the northern head-waters of the river, and Coffee Spring the southernmost spring on the river, and the far-thest distance from Rose Creek Sink (Fig. 1B). In addition, Devil's Eye and Coffee springs discharge on the western bank of the river while all other springs discharge on the eastern bank.

The timing of the initial appearance of dye at Mission Spring indicates a linear flow rate of 1.6 km/day between Rose Creek Sink and Mission Spring, reflecting predomi-nately conduit flow. The arrival of dye at some springs, but not at others, indicates that the springs along the Ichetucknee River may be sourced by two or more conduit systems, and that at least one is connected to Rose Creek Sink. The nature of the conduits feeding the springs may change with flow conditions, but because dye tracing has been done only under average flow conditions, the relationship of flow conditions to distribution of con-duits is unknown. Nonetheless, at least two sources of water for the Ichetucknee springs were confirmed by Gordon (1998), who based her conclusions on differences in the mean annual chemical composition between water in six of the springs.

There is no systematic relationship between the size of the springs and their connec-tion, or lack of connection, to Rose Creek Sink (Table 1). The long term average of total discharge from all the springs is ~10.2 m^3/sec as measured at the Highway 27 bridge, lo-cated ~2 km south of Coffee Spring (Rosenau & others, 1977). The most complete meas-urement of discharge from individual springs was made by rangers at the Ichetucknee

Table 1. Discharge of selected springs along the Ichetucknee river.

Location	Rose Creek Sink Connection*	Discharge (m^3/sec)
Ichetucknee	no	0.78
Cedar	yes	N.D.#
Blue Hole	no	2.80§
Mission Group	yes	1.12
Devil's Eye	yes	1.21
Mill Pond	yes	0.46
Coffee	no	N.D.
Spring Total:	N.A.†	6.37
Hwy 27 Bridge	N.A.	6.90
Hwy 27 Bridge, Average##	N.A.	10.20

Note: Spring discharge and river discharge at Highway 27 bridge were measured November 1990. *On the basis of dye trace study. #No data. §Combined value for Cedar and Blue Hole springs. †Not applicable. ##Av-erage of 381 measurements between 1917 and 1975. From Rosenau et al. (1977).

Springs State Park in November, 1990 (Table 1). At this time, the total discharge at the Highway 27 bridge was ~6.9 m^3/sec, about 30% lower than average. The two largest springs are Blue Hole (not sampled in this study because of limited access) and Devil's Eye. Visual estimates of the size of Cedar Spring suggest that it contributes only a small fraction of the combined discharge from Blue Hole and Cedar springs. Mission Spring is a group of closely spaced springs that feeds a single spring run. Visual estimates of Coffee Spring discharge indicate that it is the smallest of all the springs.

3 METHODS

Samples were collected from six springs roughly every two weeks from May 1996 through May 1997. The springs that were sampled, from north to south along the river, include Ichetucknee, Cedar, Devil's Eye, Mission, Mill Pond, and Coffee springs (Fig. 1B). The samples were collected from ~0.5 m below the water surface, usually within the boil of the spring. Water was preserved in three individual containers. At each sampling location, two 60 ml BOD bottles were completely filled with water, poisoned with HgCl$_3$, and sealed with ground glass stoppers using vacuum grease. One of these samples was used to measure dissolved inorganic carbon (DIC) and the other was used to measure alkalinity. A third sample was collected in 125 ml HDPE bottles, and was used to measure Cl, NO$_3$, SO$_4$, Mg, and Ca.

In addition to the biweekly samples, 43 samples were collected at higher frequency between April 23, 1997 and May 10, 1997 following a major storm. These samples were collected in a prototype automatic water sampler (Fig. 3). This autosampler uses solenoid valves mounted above and below ~100 ml PVC sample tubes. The valves, which are gas-tight to 100 psi, are opened and closed sequentially by a microprocessor connected to a rechargeable 12 V battery. The valves, sample tubes, microprocessor, and battery are all housed within a water tight pressure case, and the entire instrument is anchored to the floor of the spring, directly in the boil. When both valves sealing an individual sample tube are opened, water passively flows into the tubes, filling them in approximately one minute. The autosampler has 15 sample tubes, and the microprocessor can be programmed to collect samples at any time interval. The samples described in this paper were automatically collected at intervals ranging from 3 to 12 hours, and the entire experiment lasted 372 hours. The autosampler samples were stored only in HDPE bottles and only the concentrations of certain major ionic species (e.g. Mg, Ca, Cl, NO$_3$, and SO$_4$) were measured because gases were likely to have exchanged with the atmosphere during extraction of the samples from the sampler. This gas exchange would alter the pH, and oxygen and HCO$_3$ concentrations.

Of the samples collected biweekly, several measurements were made while in the field, including dissolved oxygen, conductivity, pH, and temperature. Oxygen was measured with an Orion model #130 portable oxygen/temperature meter and pH was measured using an Orion model #250A portable pH meter, both of which were calibrated at the start of each field day. Conductivity was measured with a YSI model #55 portable conductivity/temperature meter. The conductivity and dissolved oxygen data are missing until the fourth month of the study (August 1997) because the meters were not available until then. Temperature was measured using both the conductivity/temperature and oxygen/temperature meters while the electrodes were deployed directly in the spring boil.

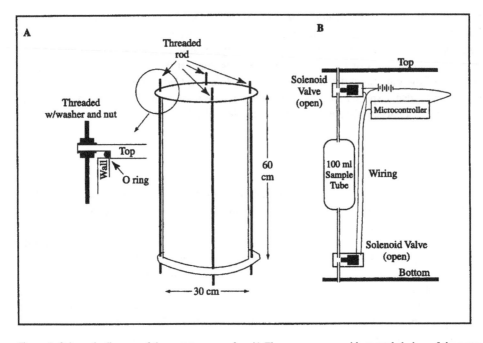

Figure 3. Schematic diagram of the prototype sampler. A) The pressure case with expanded view of the water tight seal between the case and the top and bottom (only the top is shown at expanded scale). B) Design of the interior of the sampler. For simplicity, only one sample tube and set of valves are shown. The remaining 14 sample tubes and valves have identical configurations, but all are powered and actuated by a single battery and microcontroller. The sample tubes, battery, controller, and valves are mounted on a frame, which is not shown.

The concentrations of Mg, Ca, Cl, NO_3, and SO_4 were measured using a Dionex model 500 ion chromatograph in the Department of Geology at the University of Florida. The concentrations of NO_3 reported here are in mg of NO_3/kg of water. All samples and standards were measured in triplicate, and the reported values are the average of all three measurements. Precision was estimated by measuring an internal standard following every fourth sample. This standard was a single large aliquot of water collected from Cedar Spring during the first sampling expedition on May 10, 1996. The overall precision of the measurements was calculated as the coefficient of variation, which is defined as the standard deviation of all measurements divided by their arithmetic mean for the concentration of each component in the internal standard (Table 2). Measurements were generally made on the ion chromatograph only once every one to two months, and thus in order to correct for drift between the monthly measurements, a normalizing factor was calculated as the mean of all of the internal standards divided by the mean of the internal standards within an individual run. The sample measurements in each run were multiplied by this normalizing factor, which varied at most 8% from the average value, but usually less than 2%.

Alkalinity was measured by titration with 0.1N hydrochloric acid and calculated using the Gran technique (e.g. Drever, 1988; Gieskes et al., 1992). Dissolved inorganic carbon was measured with a Coulometrics CO_2 coulometer using a 3% silver nitrate scrubber solution, nitrogen as the carrier gas, and 2N HCl to evolve the CO_2 from the water.

Table 2. Precision of Ion chromatograph measurements.

Ion	Coefficient of variation (%)*	n
Cl	0.71	77
NO_3	0.69	77
SO_4	0.49	77
Mg	1.85	57
Ca	1.63	57

*Coefficient of variation is standard deviation divided by average of all measurements of internal standards.

Both alkalinity and DIC samples were measured as soon as possible following return to the laboratory, usually within a day. Standards measured with the samples indicate the precision of these measurements to be 0.5%.

Daily precipitation totals for the region were obtained from the Suwannee River Water Management District. Data from two rain gauge stations in the area are available, with one located at Lake City and the other at Fort White (Fig. 1B). The evapotranspiration in the region was estimated using the Thornthwaite method, which is especially suited to humid climates (Thornthwaite, 1948). This method uses variables of mean air temperature, latitude, and length of day in order to estimate evapotranspiration for individual months. The mean air temperatures were obtained from the Agronomy Department at the University of Florida. The estimated evapotranspiration was subtracted from values of precipitation in order to calculate the potential recharge to the aquifer.

4 RESULTS

4.1 *Precipitation and potential recharge*

Rain gauges at Lake City and Fort White show significantly different amounts of rainfall for individual storms, although they are separated by only ~20 km (Fig. 4). These differences reflect the local nature of rainfall, particularly during summer thunderstorms. The Lake City rain gauge is located to the northeast of the river in a region with a higher potentiometric surface than at the Fort White rain gauge (Sprinkle, 1989), which is located down gradient from the Ichetucknee springs (Fig. 1). For this reason, the rainfall measured with the Fort White rain gauge is unlikely to contribute to discharge from the springs, and thus, only precipitation data from the Lake City rain gauge is used in this study.

As is typical of the region, a significant proportion of the rain fell during the rainy season months of June, July, and August (Table 3). At Lake City, precipitation was slightly below average during the 1995-96 hydrologic year (138.9 cm), and slightly above average during the 1996-97 hydrologic year (167.9 cm), the time of the study. During the 95/96 and 96/97 hydrologic years, approximately 40% and 29% of the rain fell during June, July, and August, respectively. The fraction would have been higher during the 96/97 hydrologic year except that two major storms are excluded, one that occurred at the end of May and the other at the beginning of September. If these storms are included within the rainy season, the fraction would increase to 37% of the yearly precipitation.

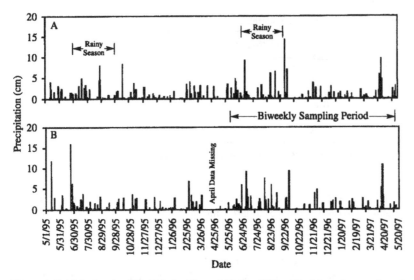

Figure 4. Precipitation data for A) Lake City and B) Fort White. The Ichetucknee springs, located at a lower potentiometric surface than the Lake City rain gauge, were sampled between May 1996 and May 1997. The rainy season is considered to include summer months of June, July, and August, although summer storms can occur in late May and early September.

Table 3. Rainfall averages.

Hydrologic year	Season/Total	Lake City (cm)	Fort White (cm)
1995/1996	Rainy Season	55.5	59.5
	Dry Season	83.4	78.4*
	Total	138.9	137.9*
1996/1997	Rainy Season	45.5	67.8
	Dry Season	122.4	100.4
	Total	167.9	168.2

Note: Precipitation data provided by Suwannee River Water Management District. *April 1996 data are missing.

During the time of this study, major storms also occurred in October 1996 and April 1997, increasing the amount of rain that typically falls during the dry season (Fig. 4).

Although a significant proportion of rain occurs during the summer months, much of this precipitation is lost as evapotranspiration. During the study year, the majority of the recharge occurred during the dry winter months (Fig. 5). Between June 1996 and May 1997, the Lake City rain gauge received 159 cm of rain, with an estimated 107 cm of water lost to evapotranspiration. Of the 52 cm of precipitation that was not lost to evapotranspiration, only 5.6 cm (~10%) recharged the aquifer during the summer months. Most of the water appears to have recharged the aquifer during the two major storms in October 1996 and April 1997. Because of these storms, these two months have a combined recharge of 39.2 cm, which represents approximately 75% of the total recharge during the year.

4.2 *Seasonal chemical variability*

There are no major changes in any of the solute concentrations over the annual sampling scale at any of the Ichetucknee springs, although there are many subtle features in the changes in chemistry through time (Fig. 6 and Table 4). In general, these changes differ among the two groups of springs that previously had been shown to be connected to Rose Creek Sink (Mission, Devil's Eye, and Mill Pond springs), and those springs where there was no detected connection (Ichetucknee, Cedar, and Coffee springs). Therefore, in the following discussion, the six springs that were sampled during this study are split into two groups on the basis of their apparent connection or lack of connection to Rose Creek Sink.

Temperature shows slight, but significant, differences among the two groups through the year, but the most pronounced seasonal changes in chemistry occur in the concentrations of oxygen, Cl, and SO_4. At Ichetucknee, Cedar, and Coffee springs, temperature is systematically colder during the winter months, by 0.1 to 0.2°C (Fig. 6A). At Mission, Devil's Eye, and Mill Pond springs, temperature is more constant than at Ichetucknee, Cedar, and Coffee springs although there are some minor variations (Fig. 6B). In contrast with temperature, the two groups of springs show pronounced differences in the concentration of oxygen. Ichetucknee, Cedar, and Coffee springs show a strong decrease in oxy

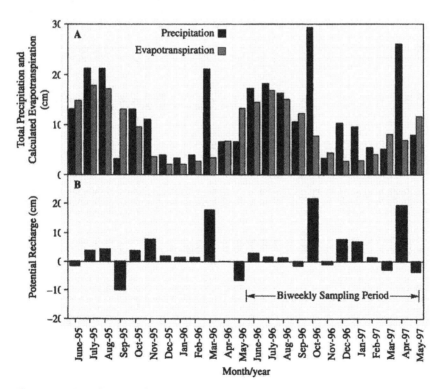

Figure 5. A) Monthly totals of precipitation and evapotranspiration calculated using the Thornthwaite method (Thornthwaite, 1948). B) Estimated amount of precipitation that is available to recharge the aquifer. In the study area, the Floridan aquifer is unconfined, and the lack of surface runoff allows an estimation of the potential recharge on the basis of total precipitation minus evapotranspiration.

Table 4. Concentrations and temperatures for six of the ichetucknee springs.

Spring	Date	T (°C)	Cond. (µS/cm)	DIC (mg/kg)	Alk. (mM)	pH	Oxygen (mg/L)	Cl (mg/kg)	NO$_3$ (mg/kg)	SO$_4$ (mg/kg)	Mg (mg/kg)	Ca (mg/kg)	SI$_{calcite}$*
Ichetucknee	05/10/96	N.D.†	N.D.	N.D.	N.D.	N.D.	N.D.	3.80	1.00	9.87	6.49	56.48	0.21
	05/23/96	N.D.	N.D.	38.0	2.89	7.7	N.D.	3.95	2.64	9.74	6.65	57.76	0.25
	05/31/96	N.D.	N.D.	34.3	3.06	7.7	N.D.	3.84	1.52	9.94	7.25	61.12	0.00
	06/19/96	N.D.	N.D.	35.4	3.00	7.5	N.D.	3.93	2.19	9.89	6.68	54.50	0.13
	07/07/96	N.D.	N.D.	34.5	3.12	7.6	N.D.	3.86	1.33	10.18	6.63	57.02	0.13
	07/17/96	N.D.	N.D.	38.0	3.11	7.6	N.D.	3.85	1.69	10.16	6.56	56.79	0.35
	08/07/96	21.8	319	36.8	3.14	7.8	4.4	3.88	1.70	9.62	6.85	59.23	0.32
	08/23/96	21.8	284	36.7	3.16	7.8	4.0	3.83	1.98	9.78	6.37	54.28	0.53
	09/08/96	21.9	298	36.7	3.17	8.0	3.8	3.88	1.92	9.69	6.72	56.17	0.34
	09/26/96	21.7	284	38.1	3.08	7.8	3.5	3.88	2.29	9.79	7.07	58.80	0.04
	10/24/96	21.7	288	37.4	3.13	7.5	2.8	3.82	2.18	9.70	6.47	57.92	0.14
	11/11/96	21.7	284	36.6	3.10	7.6	2.6	3.87	1.95	9.45	6.60	57.99	0.04
	11/24/96	21.6	299	37.2	3.09	7.5	2.7	3.81	1.86	9.51	6.61	58.59	-0.06
	12/20/96	21.7	288	38.3	3.09	7.4	2.5	3.92	2.20	9.61	6.81	58.64	-0.07
	01/11/97	21.7	289	38.1	3.06	7.4	2.6	3.93	1.96	9.81	6.84	58.00	0.11
	01/26/97	21.7	316	38.0	3.03	7.6	2.6	3.89	1.99	9.91	6.63	55.94	0.02
	02/09/97	21.7	302	38.2	3.10	7.5	2.6	3.96	2.16	9.85	6.52	55.56	0.23
	02/27/97	21.7	323	38.3	3.12	7.7	2.6	3.89	2.20	10.03	6.62	55.93	0.25
	03/10/97	21.7	323	38.4	3.12	7.7	2.5	3.89	1.86	9.91	6.96	59.88	0.04
	04/13/97	N.D.	287	37.4	3.22	7.5	N.D.	3.83	3.08	10.05	6.46	55.83	N.D.
	05/03/97	N.D.	N.D.	N.D.	N.D.	N.D.	N.D.	N.D.	N.D.	N.D.	N.D.	N.D.	N.D.
Cedar	05/10/96	N.D.	N.D.	N.D.	N.D.	N.D.	N.D.	4.21	2.83	6.37	6.26	56.77	0.13
	05/23/96	N.D.	N.D.	37.9	3.01	7.6	N.D.	4.18	2.15	6.62	6.56	58.55	0.13
	05/31/96	N.D.	N.D.	37.5	3.10	7.6	N.D.	4.14	1.58	6.43	6.49	57.94	0.14
	06/19/96	N.D.	N.D.	35.0	3.05	7.5	N.D.	4.18	1.95	6.43	6.31	46.74	-0.05
	07/07/96	N.D.	N.D.	35.6	3.05	7.8	N.D.	4.20	1.49	6.73	6.63	59.89	0.34
	07/17/96	N.D.	N.D.	38.5	3.22	7.5	N.D.	4.24	1.64	6.74	6.53	59.19	0.06
	08/07/96	21.7	317	36.7	3.16	7.9	3.3	4.22	1.58	6.57	6.42	58.52	0.45
	08/23/96	21.7	283	37.2	3.11	7.8	3.6	4.14	1.75	6.54	6.55	57.34	0.34
	09/08/96	21.7	282	36.8	3.19	7.6	3.2	4.38	1.46	6.70	6.21	56.71	0.14
	09/26/96	21.7	282	37.5	3.08	7.6	3.0	4.12	2.15	6.49	6.48	57.50	0.13

Date													
10/24/96	21.6	290	39.2	3.37	7.5	2.8	4.12	2.38	6.68	6.36	58.00	0.07
11/11/96	21.5	311	36.7	3.24	7.5	2.3	4.26	1.89	6.25	6.18	57.00	0.05
11/24/96	21.7	300	37.6	3.09	7.4	2.2	4.21	1.90	6.33	6.11	57.65	-0.06
12/20/96	21.5	289	38.6	3.11	7.6	2.4	4.18	1.84	6.48	6.49	58.43	0.14
01/11/97	21.5	289	38.7	3.07	7.4	2.2	4.16	2.04	6.60	6.72	58.24	-0.06
01/26/97	21.6	299	38.1	3.14	7.7	2.3	4.21	1.63	6.60	6.21	56.60	0.24
02/09/97	21.6	301	38.5	3.09	7.4	2.3	4.20	2.16	6.69	6.38	56.59	-0.07
02/27/97	21.7	313	38.2	3.07	7.7	2.2	4.25	1.96	6.32	6.28	58.48	0.24
03/10/97	21.6	323	38.8	3.21	7.6	2.2	4.20	2.15	6.62	6.56	59.64	0.16
04/13/97	21.6	291	37.3	3.15	7.5	2.3	4.13	3.32	6.89	6.49	59.18	0.05
05/03/97	21.7	321	38.4	3.15	N.D.	2.4	4.44	2.65	6.89	6.48	59.53	N.D.
05/10/96	N.D.	N.D.	N.D.	N.D.	N.D.	N.D.	N.D.	N.D.	N.D.	N.D.	N.D.	N.D.
05/23/96	N.D.	N.D.	30.7	2.37	7.7	N.D.	3.27	1.96	11.67	4.86	50.57	0.08
05/31/96	N.D.	N.D.	30.8	2.60	7.7	N.D.	3.26	1.99	11.54	4.89	49.98	0.11
06/19/96	N.D.	N.D.	30.5	2.60	7.7	N.D.	3.27	1.90	11.62	5.19	51.94	0.13
07/07/96	N.D.	N.D.	30.6	2.67	7.8	N.D.	3.33	1.10	11.73	5.23	51.57	0.23
07/17/96	N.D.	N.D.	30.9	2.78	7.6	N.D.	3.23	1.24	10.94	4.78	51.42	0.05
08/07/96	21.8	268	30.3	2.60	7.8	2.5	3.27	2.22	11.22	4.45	50.62	0.22
08/23/96	21.8	245	30.2	2.60	7.8	2.5	3.25	2.10	11.38	4.59	49.46	0.21
09/08/96	21.8	248	30.3	2.66	7.7	2.3	3.35	1.78	11.54	4.78	50.20	0.12
09/26/96	21.8	248	31.7	2.64	7.7	2.0	3.23	1.66	11.71	4.77	50.50	0.12
10/24/96	21.7	235	N.D.	N.D.	7.7	1.7	N.D.	N.D.	N.D.	N.D.	N.D.	N.D.
11/11/96	21.8	259	29.4	2.62	7.6	1.7	3.29	1.92	11.08	4.54	50.00	0.02
11/24/96	21.7	251	30.5	2.59	7.6	1.7	3.27	1.99	11.35	4.47	50.44	0.01
12/20/96	21.7	254	31.9	2.66	7.6	1.7	3.29	2.08	11.46	4.91	50.68	0.03
01/11/97	21.7	264	31.8	2.64	7.7	1.6	3.27	1.86	11.55	5.03	51.49	0.13
01/26/97	21.7	265	32.1	2.72	7.7	1.6	3.26	2.00	11.58	5.11	50.04	0.13
02/09/97	21.8	282	32.1	2.79	7.7	1.6	3.31	1.37	11.56	4.96	54.83	0.18
02/27/97	21.7	283	32.5	2.67	7.7	1.6	3.26	1.53	11.47	4.99	55.49	0.16
03/10/97	21.7	251	32.3	2.85	7.6	1.7	3.25	1.44	11.55	5.25	53.73	0.08
04/13/97	21.7	276	31.9	2.63	7.7	1.8	3.20	2.11	11.15	4.90	52.09	0.13
05/03/97	N.D.	N.D.	32.2	2.59	N.D.	N.D.	3.25	1.60	11.05	4.76	51.29	N.D.

Coffee

Coffee

Table 4. Continued.

Spring	Date	T (°C)	Cond. (µS/cm)	DIC (mg/kg)	Alk. (mM)	pH	Oxygen (mg/L)	Cl (mg/kg)	NO$_3$ (mg/kg)	SO$_4$ (mg/kg)	Mg (mg/kg)	Ca (mg/kg)	SI$_{calcite}$*
Mission	05/10/96	N.D.	N.D.	N.D.	N.D.	N.D.	N.D.	6.05	0.96	11.30	7.90	52.97	N.D.
	05/23/96	N.D.	N.D.	35.9	2.92	7.6	N.D.	6.02	1.97	10.86	7.31	49.65	0.05
	05/31/96	N.D.	N.D.	34.9	2.95	7.7	N.D.	5.84	0.42	10.50	8.24	56.00	0.20
	06/19/96	N.D.	N.D.	N.D.	N.D.	7.6	N.D.	N.D.	N.D.	N.D.	N.D.	N.D.	N.D.
	07/07/96	N.D.	N.D.	35.2	3.08	7.6	N.D.	5.81	1.72	10.63	7.87	53.22	0.10
	07/17/96	N.D.	N.D.	36.0	2.98	7.9	N.D.	5.84	0.00	10.21	7.93	54.01	0.39
	08/07/96	21.7	307	34.1	3.06	7.7	0.2	5.80	0.00	10.95	7.81	53.11	0.20
	08/23/96	21.7	278	34.1	2.96	7.8	0.3	5.75	0.56	11.08	7.59	51.57	0.27
	09/08/96	21.7	279	35.0	3.01	7.7	0.3	5.76	0.44	11.27	8.09	52.17	0.18
	09/26/96	21.7	280	35.4	2.96	7.6	0.4	5.83	0.64	11.13	8.16	53.12	0.08
	10/24/96	21.7	279	33.8	2.84	7.6	0.3	5.47	1.22	10.08	6.88	48.82	0.04
	11/11/96	21.7	293	32.6	2.92	7.5	0.2	5.62	0.00	10.52	7.13	49.04	-0.05
	11/24/96	21.6	284	34.0	2.82	7.6	0.2	5.56	0.00	10.53	7.39	51.10	0.05
	12/20/96	21.7	274	35.2	2.99	7.6	0.2	5.68	0.61	10.00	7.85	54.10	0.09
	01/11/97	21.7	276	35.4	2.83	7.7	0.3	5.66	1.24	9.87	7.46	53.21	0.17
	01/26/97	21.7	287	35.0	3.00	7.7	0.2	5.74	0.40	9.88	7.27	50.50	0.17
	02/09/97	21.7	286	35.1	2.91	7.7	0.2	5.79	0.10	9.75	7.43	50.85	0.16
	02/27/97	21.7	306	35.3	2.87	7.8	0.3	5.74	0.15	9.65	7.51	60.60	0.32
	03/10/97	21.7	307	35.4	3.00	7.5	0.2	5.69	0.55	9.60	7.50	52.18	-0.02
	04/13/97	21.7	275	34.9	2.89	7.5	0.3	5.69	2.22	9.24	7.32	51.52	-0.04
	05/03/97	N.D.	301	35.3	2.81	N.D.	0.4	5.67	1.66	8.80	6.99	49.44	N.D.
Devil's Eye	05/10/97	N.D.	N.D.	N.D.	N.D.	N.D.	N.D.	N.D.	N.D.	N.D.	N.D.	N.D.	N.D.
	05/23/96	N.D.	N.D.	35.9	2.67	7.6	N.D.	6.54	1.73	20.89	9.89	54.28	0.04
	05/31/96	N.D.	N.D.	36.0	3.04	7.7	N.D.	6.65	0.70	20.41	9.89	54.04	0.19
	06/19/96	N.D.	N.D.	N.D.	N.D.	7.8	N.D.	N.D.	N.D.	N.D.	N.D.	N.D.	N.D.
	07/07/96	N.D.	N.D.	34.9	3.03	7.6	N.D.	5.80	0.19	11.13	7.52	53.06	0.09
	07/17/96	N.D.	N.D.	36.2	2.99	7.7	N.D.	6.53	0.66	20.24	10.01	54.55	0.19
	08/07/96	21.8	330	35.6	2.98	7.8	0.3	6.56	0.16	22.07	10.23	55.11	0.29
	08/23/96	21.8	299	35.9	3.03	7.6	0.3	6.65	0.27	22.01	9.94	53.13	0.09
	09/08/96	21.8	297	34.8	3.05	7.6	0.3	6.58	0.52	21.89	10.35	53.51	0.09
	09/26/96	21.7	301	35.0	3.17	7.8	0.4	6.57	0.19	21.40	9.72	53.91	0.31
	10/24/96	21.8	294	34.3	2.90	7.7	0.3	6.08	0.60	16.83	8.44	50.77	0.15

	Date												SI
	11/11/96	21.8	320	33.7	2.89	7.5	0.2	6.41	0.89	21.59	9.84	53.38	-0.03
	11/24/96	21.7	294	34.4	2.90	7.5	0.2	6.42	0.69	21.35	9.60	52.42	-0.04
	12/20/96	21.8	296	36.0	2.90	7.4	0.2	6.41	0.76	19.89	9.56	53.04	-0.13
	01/11/97	21.8	297	36.1	2.95	7.7	0.2	6.40	1.76	19.66	9.57	52.61	0.17
	01/26/97	21.8	308	35.9	2.93	7.7	0.3	6.37	1.55	19.26	9.25	52.22	0.17
	02/09/97	21.8	308	36.0	2.90	7.6	0.2	6.49	1.42	19.45	9.64	52.79	0.07
	02/27/97	21.8	329	36.3	3.00	7.7	0.3	6.51	0.99	19.35	9.14	50.62	0.16
	03/10/97	21.8	329	36.0	2.96	7.5	0.3	6.35	0.96	18.95	9.64	53.29	-0.02
	04/13/97	21.8	295	35.9	3.04	N.D.	0.3	6.16	1.52	17.42	9.28	51.98	N.D.
	05/03/97	N.D.	323	36.4	2.90	N.D.	0.5	6.65	1.43	17.01	9.39	53.48	N.D.
Mill Pond	05/10/96	N.D.	N.D.	N.D.	N.D.	N.D.	N.D.	8.32	0.00	41.95	11.80	57.61	N.D.
Mill Pond	05/23/96	N.D.	N.D.	35.7	2.82	7.6	N.D.	8.39	0.45	42.29	11.95	58.05	0.08
	05/31/96	N.D.	N.D.	36.1	2.85	7.6	N.D.	8.41	0.00	41.18	12.00	57.84	0.08
	06/19/96	N.D.	N.D.	N.D.	2.88	7.6	N.D.	8.32	1.18	40.45	11.55	55.73	0.07
	07/07/96	N.D.	N.D.	35.3	2.91	7.6	N.D.	8.18	0.31	39.56	11.24	54.39	0.07
	07/17/96	N.D.	N.D.	35.8	2.98	7.5	N.D.	8.41	0.00	41.13	11.99	58.29	0.00
	08/07/96	21.9	370	35.4	2.96	7.7	0.3	8.56	0.00	43.67	12.02	57.57	0.20
	08/23/96	21.9	333	35.5	2.99	7.5	0.3	8.39	0.07	43.82	11.31	56.43	-0.01
	09/08/96	21.9	333	35.1	3.01	7.7	0.4	8.43	0.77	43.55	12.27	58.72	0.21
	09/26/96	22.0	332	36.2	3.03	7.6	0.4	8.41	0.65	42.84	12.16	57.80	0.11
	10/24/96	21.9	328	35.2	2.95	7.6	0.2	8.32	0.00	42.29	11.63	57.17	0.09
	11/11/96	21.9	360	33.5	2.91	7.5	0.2	8.41	0.25	43.60	11.67	57.81	-0.01
	11/24/96	21.9	346	34.6	2.90	7.4	0.2	8.30	0.14	43.29	11.39	57.31	-0.11
	12/20/96	21.8	328	35.6	2.89	7.6	0.3	8.24	0.33	40.84	11.60	56.61	0.08
	01/11/97	21.8	328	35.4	2.89	7.5	0.2	8.11	0.32	40.16	11.38	65.35	0.04
	01/26/97	21.9	340	35.7	2.92	7.7	0.3	8.04	0.11	39.70	11.24	56.07	0.18
	02/09/97	21.9	340	35.8	2.93	7.5	0.3	8.10	0.25	39.65	11.38	54.64	-0.03
	02/27/97	21.9	363	36.1	3.04	7.8	0.3	7.94	0.32	38.99	11.47	56.28	0.30
	03/10/97	21.9	363	36.0	2.93	7.6	0.3	7.89	0.00	39.13	11.75	58.13	0.10
	04/13/97	21.9	362	35.7	2.93	7.6	0.3	7.79	1.51	38.38	12.24	60.76	0.11
	05/03/97	21.9	353	36.0	2.88	7.5	0.4	7.86	0.64	37.26	11.16	56.53	-0.02

* SI = Saturation with respect to calcite; calculated as log(IAP/K$_{calcite}$). Activity coefficients calculated from Debye-Hückel equation.
†No data.

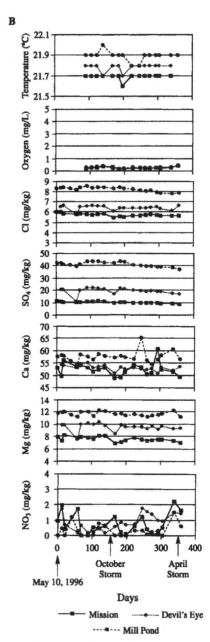

Figure 6. Temperature and concentrations of oxygen, Cl, SO₄, Ca, Mg, and NO₃ for six of the Ichetucknee springs for one full year starting May 10, 1996. A) Ichetucknee, Cedar, and Coffee springs. These springs showed no apparent connection with Rose Creek Sink during previous dye trace studies. B) Mission, Devil's Eye, and Mill Pond springs. These springs were shown to be connected to Rose Creek Sink by dye tracing. The times are marked with an arrow for two major storms that occurred during October 1996 and April 1997.

gen concentrations through the fall, with the lowest oxygen concentrations occurring in the winter and spring months. Mission, Devil's Eye, and Mill Pond springs show low oxygen concentrations throughout the year, and although there is some variation, the overall low concentrations suggest the seasonal variations may not be significant (Table 4). Among the two groups of springs the Cl and SO_4 concentrations show opposite trends to temperature and oxygen concentrations. At Ichetucknee, Cedar, and Coffee springs, the Cl and SO_4 concentrations vary little throughout the year. At Mission, Devil's Eye, and Mill Pond springs, however, the SO_4 concentrations decrease throughout the year. This decrease is most pronounced at Mill Pond Spring, because it is characterized by the highest SO_4 concentration of all the sampled springs. Mill Pond Spring also shows the largest decrease in Cl concentration through the year of all the springs.

Differences in the seasonal concentrations of Cl and SO_4, as well as temperature, are clearly illustrated among the two groups of springs by comparing their coefficients of variation for all of the biweekly measurements (Table 5). The oxygen concentrations vary by only ~0.25 mg/L at Mission, Devil's Eye, and Mill Pond springs, and low concentrations cause their coefficients of variation to be larger than those at Ichetucknee, Cedar, and Coffee springs. Coefficients of variation for temperature, as well as Cl and SO_4 concentrations, also reflect the differences between the two groups of springs that are shown in Figure 6. The largest variations in temperature through the year occur at Ichetucknee, Cedar, and Coffee springs, while the largest variations in Cl and SO_4 concentrations occur at Mission, Devil's Eye, and Mill Pond springs (Table 5).

There is little systematic variation in other solutes that were measured (e.g. Ca, Mg, pH, DIC, NO_3, and alkalinity) through the year at any of the springs (Fig. 6 and Table 4). Calcium concentrations are more variable than Mg concentrations, but there is no systematic trend through time in either of their concentrations (Fig. 6). The only major change from the background value in Mg concentrations occurs at Mission and Devil's Eye springs following the major storm in October 1996 (Fig. 6). At Mission Spring, the Mg concentration decreased by ~16% from the prestorm value, while at Devil's Eye Spring, the Mg concentration decreased by ~14% from the prestorm value. No high frequency samples were collected following this storm, however, and thus there is no detailed record of how the storm affected concentrations. None of the sampled springs showed a similar decrease following the major storm in April 1997 when high frequency samples were collected.

The largest range in concentration of all the solutes occurs in the concentration of NO_3, but similar to most of the solutes, there is no systematic difference in concentration

Table 5. Coefficient of variation of biweekly measurements.

Spring	T (%)	Oxygen (%)	Cl (%)	NO_3 (%)	SO_4 (%)	Mg (%)	Ca (%)
Ichetucknee	0.33	21.65	1.19	22.37	2.02	3.27	3.11
Cedar	0.37	18.61	1.87	23.12	2.68	2.52	4.66
Coffee	0.24	18.60	1.06	18.08	2.06	4.96	3.22
Mission	0.12	24.18	3.38	94.62	6.80	5.09	5.15
Devil's Eye	0.17	24.59	2.66	59.69	13.32	6.78	2.17
Mill Pond	0.21	26.59	2.37	117.44	4.74	2.98	3.94

during the summer rainy season and the remainder of the year. As shown by the coefficient of variation, the largest variation in NO_3 concentration occurs at Mission, Devil's Eye, and Mill Pond springs, which have overall lower concentrations than Ichetucknee, Cedar, and Coffee springs (Table 5). Much of the variation in NO_3 concentrations of the springs occurs at the beginning of the sampling period, and all of the springs show a sharp increase in NO_3 concentrations at the end of the sampling period. These samples were collected immediately prior to the large storm at the end of April 1997, but similar variations are not observed prior to the other large storm in October 1996 (Fig. 4).

4.3 *Storm response at Mill Pond Spring*

In the seven days between April 24 and April 30, 23.7 cm of rain fell at the Lake City rain gauge. In comparison, a total of only 8.9 cm of rain fell in the 60 days prior to this storm (Fig. 7). The storm started early on April 24, with most of the rain falling between 6 and 8 AM. The autosampler was installed at Mill Pond Spring at 5:25 PM on the afternoon of April 24, approximately 10 hours after the start of the storm, in order to observe variations in the chemistry of the discharge water that might occur as a result of the large amount of rain following a prolonged dry period.

The variations in concentrations in these water samples show a pattern similar to the long-term, seasonal samples. In particular, Cl and SO_4 decrease in concentration from an initial high value following the storm (Table 6 and Fig. 8). Most of the samples after the start of the storm have Cl concentrations that are significantly lower than the yearly average (Fig. 8A). The highest concentrations of Cl occur immediately after the start of the storm, and exceed the concentrations immediately prior to the storm by ~0.3 mg/kg. These elevated Cl concentrations are around the average concentration for the preceding year. The sample with the highest SO_4 concentration is significantly lower than the yearly average of 41.1 mg/kg, however, while the overall decrease in SO_4 concentration with time is similar to the decrease in concentration observed over the preceding year (Figs 6 and 8B). In contrast, the Mg and Ca concentrations show some variation, but no defini-

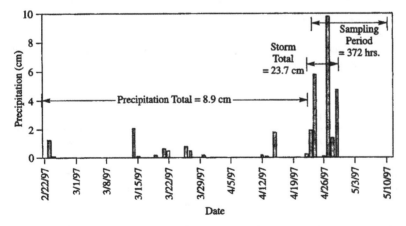

Figure 7. Daily precipitation record for two months preceeding and during the April 1997 storm. There was only 8.9 cm of rain during the two months prior to the storm, but 23.7 cm fell during the seven days of the storm. High frequency sampling started on the evening of April 24 and continued through May 10.

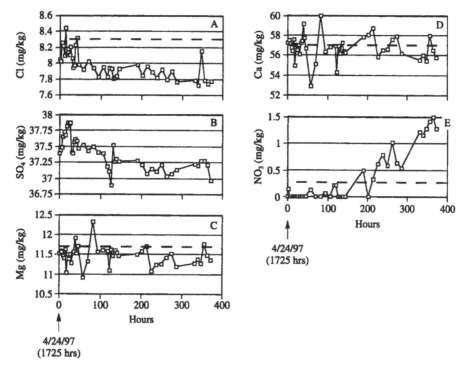

Figure 8. Concentrations of various solutes at Mill Pond Spring over 372 hours following the initiation of the April 1997 storm. A) Cl concentrations, B) SO$_4$ concentrations, C) Mg concentrations, D) Ca concentrations, E) NO$_3$ concentrations. The dashed lines represent the average of each concentration over the proceeding year (e.g. Fig. 6). The average concentration of SO$_4$ is 41.1 mg/kg and not plotted on the figure.

tive increasing or decreasing trend in the change in concentration following the storm (Figs 8C and D). Both Ca and Mg concentrations vary around the long term average, and the range in this variation is only slightly smaller than the range observed over the entire year. The largest variations in Ca and Mg concentrations occur between the peak periods of rain on April 25 and 28.

The largest change in concentration of any of the solutes occurs in the NO$_3$ concentrations (Fig. 8E). During the first ~150 hours of sampling most of the samples have no detectable NO$_3$. Nitrate concentrations start to increase ~200 hours after the start of the storm, and rapidly reach the highest measured value of 1.5 mg/kg about 170 hours later. The highest NO$_3$ concentration measured is significantly greater than the average NO$_3$ concentration for the previous year, but is similar to the value of a sample that had been collected approximately 2 weeks prior to the storm (Table 4). At the time sampling was stopped, 372 hours after the start of the storm, there were no indications from field measurements (e.g. oxygen, pH, conductivity, or temperature) that concentrations of any of the solutes had changed greatly. For this reason, the high frequency sampling was ended, although it would have been important to observe when the NO$_3$ concentrations declined to prestorm values.

Table 6. Concentrations of solutes from autosampler samples.

Date	Time	Hours*	Cl (mg/kg)	NO$_3$ (mg/kg)	SO$_4$ (mg/kg)	Mg (mg/kg)	Ca (mg/kg)
4/24/97	1725	0.00	8.04	0.00	37.40	11.53	57.26
	2025	3.00	8.03	0.14	37.45	11.59	57.52
	2325	6.00	8.26	0.00	37.48	11.57	57.16
4/25/97	225	9.00	8.22	0.00	37.65	11.47	56.82
	525	12.00	8.09	0.00	37.73	11.42	56.47
	825	15.00	8.44	0.00	37.68	11.64	57.59
	1125	18.00	8.11	0.00	37.82	11.05	54.93
	1425	21.00	8.13	0.00	37.86	11.50	57.07
	1725	24.00	8.11	0.00	37.86	11.38	56.52
	2025	27.00	8.21	0.00	37.87	11.50	57.01
	2325	30.00	8.07	0.00	37.41	11.27	56.09
4/26/97	225	33.00	7.94	0.00	37.40	11.53	57.28
	525	36.00	7.97	0.00	37.58	11.58	57.46
	825	39.00	8.23	0.00	37.61	11.91	59.16
	1125	42.00	8.32	0.00	37.59	11.53	57.80
	1510	45.75	7.98	0.00	37.47	11.72	56.72
4/27/97	310	57.75	7.92	0.12	37.52	10.91	52.92
	1510	69.75	8.02	0.00	37.43	11.33	55.14
4/28/97	310	81.75	7.94	0.00	37.49	12.33	59.98
	1510	93.75	7.82	0.07	37.40	11.57	56.38
4/29/97	310	105.75	7.95	0.00	37.39	11.62	56.87
	1400	116.58	7.83	0.22	37.19	11.59	56.91
	1830	121.08	7.94	0.22	37.11	11.10	54.31
	2300	125.58	7.93	0.00	36.90	11.62	56.97
4/30/97	330	130.08	7.81	0.00	37.53	11.47	56.17
	800	134.58	7.82	0.00	37.26	11.60	57.24
	1230	139.08	7.84	0.00	37.31	11.55	56.47
	1700	143.58	7.93	0.00	37.27	11.48	56.32
5/2/97	1530	190.08	7.99	0.49	37.27	11.51	57.84
5/3/97	330	202.08	7.85	0.00	37.21	11.59	58.12
	1530	214.08	7.96	0.33	37.07	11.71	58.75
5/4/97	330	226.08	7.89	0.62	37.15	11.07	55.91
	1530	238.08	7.82	0.79	37.12	11.24	56.53
5/5/97	330	250.08	7.92	0.59	37.22	11.27	56.58
	1530	262.08	7.80	1.03	37.03	11.44	57.59
5/6/97	330	274.08	7.90	0.63	37.07	11.53	57.89
	1530	286.08	7.77	0.53	37.14	11.21	56.17
5/8/97	1350	332.42	7.79	1.22	37.23	11.28	55.58
	2150	340.42	7.73	1.16	37.21	11.39	56.03
5/9/97	550	348.42	8.15	1.27	37.28	11.28	55.48
	1350	356.42	7.79	1.41	37.28	11.77	57.99
	2150	364.42	7.74	1.50	37.22	11.49	56.56
5/10/97	550	372.42	7.78	1.29	36.97	11.36	55.80

*Time elapsed since the start of sampling.

5 DISCUSSION

5.1 *Diffuse versus conduit flow*

Of critical importance to subsurface flow in karst aquifers, where porosity and permeability vary widely, is the volume of water that flows through large conduits, such as caves, and the volume of diffuse flow through intergranular pore spaces and fractures. The relative proportions of diffuse and conduit flow have previously been identified and quantified on the basis of chemographs of chemical composition and temperature of spring discharge and their coefficients of variation (e.g. Shuster & White, 1971; Ternan, 1972). Springs with high coefficients of variation have large and systematic seasonal variations in temperature and hardness (calculated as $CaCO_3$). These springs are believed to be have a source controlled by conduit flow because of rapid infiltration of surface water and short residence times in the subsurface, which limit equilibration with the wall rock (e.g. Shuster & White, 1971; Ternan, 1972; Smart & Ford, 1986). Springs characterized by low coefficients of variation are commonly close to saturation with respect to calcite or dolomite. These springs are believed to have a source from intergranular porosity rather than conduits, reflecting long residence time that allows equilibration with the wall rocks (Shuster & White, 1971; Ternan, 1972; Dreiss, 1989a). The values of coefficients of variation have been found to correlate with calculated flow through times from recharge water for springs in the Central Pennines of England, providing a technique for determining flow through time on the basis of the coefficient of variation (Ternan, 1972).

Water discharging from the Floridan aquifer appears to have long residence times on the basis of its chemical compositions. These long residence times are also seen as low coefficients of variation and saturation with calcite, which reflect equilibration with the wall rock (Table 5). The coefficients of variation shown in Table 5, when applied to Ternan's (1972) correlation of flow through times and coefficient of variation, indicate flow through times for the Ichetucknee springs of between 40 and 200 days (Table 5). Ternan's (1972) relationship may not be directly applicable to the Floridan aquifer springs, however, because of differences between the karst aquifers in the Pennines and Florida carbonate platform. Additional evidence for long residence times was shown by Katz and Hornsby (1988) who found on the basis of CFC concentrations that water discharging from Little River Spring, located ~20 km west of the Ichetucknee springs, has an average age of 12 to 25 yrs. These results suggest that the source of the Ichetucknee springs may be controlled by diffuse flow.

Regardless of the chemical evidence for diffuse flow, however, conduits are an important characteristic of the aquifer in the region. Cave diving exploration has shown long and large open caverns throughout the area, and flow rates on the order of kilometers per day within the Ichetucknee basin indicate these caverns are used as flow paths. This apparent discrepancy between the presence of conduits, but chemical indicators of diffuse flow, may reflect fundamental differences in the characteristics of the Floridan aquifer from other previously studied karst aquifers. In particular, where the Floridan aquifer is unconfined, the ground water table is located at most only a few meters below the ground surface because of the flat topography and high rainfall. As a result of the shallow water table, most conduits are continuously filled with water. Recharge water when it infiltrates the aquifer would mix with large volumes of water as it flows through filled conduits.

These large volumes of water in the conduits are likely to buffer rapid changes in the composition of the spring discharge water following storms and through seasons.

5.2 *Seasonal controls of water chemistry*

On the basis of dye trace studies, the six springs sampled during this study can be separated into two different groups. Within each group, there are consistent variations in the chemical composition through time, although these variations differ between the groups. These differences are clearly exemplified by the Cl and SO_4 concentrations. Rain water in the region has Cl and SO_4 concentrations that are approximately 1 to 2 orders of magnitude less than the ground water (Gordon, 1998), and thus infiltration of rain water should decrease the concentrations of both solutes. The Cl and SO_4 concentrations at Ichetucknee, Cedar, and Coffee springs, however, show little seasonal variation, implying that recharge water completely mixes with the ground water prior to discharge. This result supports the dye trace study, which indicates the springs may not have a direct connection to swallets (Fig. 1B). In contrast, Cl and SO_4 concentrations decrease through time at Mission, Devil's Eye, and Mill Pond springs. For example, the SO_4 and Cl concentrations decrease by ~15% and 9% respectively at Mill Pond Spring between August 1996 and May 1997. This decrease may result from dilution resulting from point recharge into Rose Creek Sink and possibly other swallets such as Cannon Creek and Clay Hole Creek Sink (Fig. 1). The concentrations decrease smoothly through time, however, even following the October 1996 and April 1997 storms, which generated large amounts of recharge (Figs 5 and 6). The smooth decrease supports the earlier conclusion that recharge water mixes with large volumes of ground water, either in intergranular porosity, or in water filled conduits.

Although the Cl and SO_4 concentrations imply there is a greater influence of surface water mixing with water discharged from Mission, Devil's Eye, and Mill Pond springs than from Ichetucknee, Cedar, and Coffee springs, the oxygen and temperature data do not reflect this mixing. Gordon (1998) suggested that the low oxygen content and elevated temperatures may reflect a deeper source and longer residence time in the subsurface for the discharging water. A deep source would allow more uniform equilibration of temperature with the ground water and aquifer rocks and longer residence times would allow greater microbial oxygen utilization, thereby lowering its concentration. Thus, those springs with apparent connections to Rose Creek Sink (e.g. Mission, Devil's Eye, and Mill Pond) may have a source from greater depth than the other three springs.

5.3 *Storm controls of water chemistry*

Chemical variations following storms have been used to identify the timing of water released from subsurface conduit systems and to identify storage locations of water in the subsurface (e.g. Smart & Ford, 1986; Hess & White, 1988; Dreiss, 1989a;b; Lakey & Krothe, 1996). For example, the proportions of water that are derived from phreatic and vadose conduits, epikarst, and soils were separated on the basis of discharge records, δD and $\delta^{18}O$ values, and solute chemistry (Lakey & Krothe, 1996). Chemographs of Ca and Mg concentrations, as well as hydrographs of spring discharge, allow separation of old and new water, where old water has been stored in the aquifer for sufficiently long periods of time to equilibrate with the wall rocks of the aquifer, and new water has more recently recharged the aquifer following storms (Dreiss, 1989b). Most springs that have been studied show significant and rapid responses, both in chemical composition and discharge, to storm water recharge. In these springs, concentrations of various solutes can vary by more than 50% following large storms (e.g. Dreiss, 1989b; Lakey & Krothe, 1996). The absolute response varies, however, depending on characteristics of the aquifer from different regions (e.g. Padilla & others, 1994).

Unlike variations in concentrations at many other karst springs, most solutes show only slight changes in concentration at Mill Pond Spring following the large storm in April 1997 (Fig. 8). Similar to the seasonal data (Fig. 6), the smallest variations occur in Ca and Mg concentrations following the storm, and the concentrations of these solutes vary around the average concentrations for the entire year. In contrast, Cl and SO_4 concentrations exhibit small but significant trends in their concentrations. The Cl concentrations decreased by ~7%, and the SO_4 concentrations decreased by ~3% following an initial high value after the storm. The initially high concentrations of Cl and SO_4 following the onset of the rain may reflect the initial pulse of water from the epikarst zone that was evaporatively concentrated during the two dry months before the storm (e.g. Lakey and Krothe, 1996). The decline in concentration of Cl and SO_4 following the storm may reflect dilution from the rain water because it has concentrations that are one to two orders of magnitude less than ground water and neither Cl nor SO_4 takes part in fluid-solid reactions. The smaller decrease in SO_4 than in Cl concentration implies that the Cl/SO_4 ratio of the water recharging the aquifer is less than that of the ground water. There are no samples of surface water runoff during the storm, but runoff, as well as precipitation, is likely to have quite variable compositions (e.g. Gordon, 1998).

The Ca and Mg concentrations must be influenced by other processes, probably fluid-solid reactions, because they do not show the same dilution effects as Cl and SO_4. Although components of the carbonate system (e.g. alkalinity and DIC) were not measured on these samples, the Ca concentrations imply that water discharged after the storm is near saturation with calcite (Table 4). If these samples are in equilibrium with calcite, the recharge water must have rapidly equilibrated with the aquifer rocks. This rapid equilibration suggests that the wall rocks are easily dissolved.

5.4 *Infiltration of contaminant* NO_3

Several observations suggest that NO_3 concentrations are controlled by contamination: a rapid increase at Mill Pond Spring following the April 1997 storm, high variation in concentration throughout the year, and lack of seasonal trends. Land use in the basin is largely agricultural, and NO_3 could have a source from fertilizer dissolved in runoff, which subsequently infiltrates the aquifer. Other possible contaminant sources could be from dairy farming (e.g. Boyer & Pasquarell, 1995; 1996; Andrews, 1994) and from septic systems. Throughout the year, the concentration of NO_3 in water at Rose Creek Sink varied below detection to 3 mg/kg, and had little correlation with seasonal variations in rainfall (Gordon, 1998). The wide range in NO_3 concentrations of this surface water implies there is contaminant NO_3 entering the sinkhole, and that NO_3 is consumed by some process, possibly by plant growth during photosynthesis. If the NO_3 is from agricultural or other sources of contaminants, it would be decoupled from the sources of Cl, SO_4, and oxygen, resulting in a lack of correlation with these solutes.

The potential for high input of NO_3 into the groundwater following storms is reflected in a sharp increase in NO_3 concentrations ~200 hours after the beginning of the April storm. This increase occurs simultaneously with a decrease in the Cl and SO_4 concentrations (Fig. 8). The length of time for NO_3 to emerge from the springs following the start of the storm also corresponds to the amount of time required for dye to travel from Rose Creek Sink to Mill Pond Spring (reported in Gordon, 1998). The correlations of change of NO_3 concentrations with Cl and SO_4 concentrations and the timing of the emergence of the dye imply that storm runoff into Rose Creek Sink may be a source of NO_3 in the spring. In order for the storm water to increase NO_3 concentrations when discharged from the springs, however, the concentration of NO_3 in the storm runoff must have been higher than that measured at the spring (e.g. Fig. 6B). The lack of NO_3 in the spring water during the first 200 hours after the start of the storm suggests that water flushed from the vadose or epikarst zones is NO_3 free. This result differs from elevated NO_3 concentration found by Iqbal & Krothe (1995) in the epikarst layer of Indiana karst. The difference in behavior of the Florida and Indiana karst may result from compositional differences in the epikarst. The Indiana epikarst contains more clay than the Florida epikarst, which is predominately sand. High sand content would cause rapid and widespread infiltration through the epikarst, and NO_3 contained in the epikarst would be diluted by the large volumes of ground water.

The potential contribution from contaminated surface water is also shown by the high variability in NO_3 concentrations throughout the year, as well as from the lack of seasonal changes (Fig. 6 and Table 5). The high variation in NO_3 concentrations in spring water probably reflects the relatively rapid variations in concentrations of NO_3 in surface water that would be expected from a combination of contamination during periods of rain and NO_3 utilization during photosynthesis. Those springs with an apparent connection to Rose Creek Sink, e.g. Mission, Devil's Eye, and Mill Pond springs, have significantly higher coefficients of variation than the other springs (Table 4), which may reflect a direct connection of these springs to areas with point source recharge. These springs have overall lower NO_3 concentrations than the remaining springs, however, possibly as a result of a deeper source of water (e.g. Gordon, 1998).

6 CONCLUSIONS

On the basis of early work (e.g. Shuster & White, 1971; Ternan, 1972; Smart & Ford, 1986; Dreiss, 1989b; Padilla & others, 1994; Lakey & Krothe, 1996), spring discharge should have small variations in composition through time if the springs are sourced from small fractures and intergranular porosity. The variations should be small regardless of changes in seasonal rainfall or following major storms, although the storms may provide a significant fraction of the recharged water. Although dye trace studies and cave diving exploration show that conduits provide important sources for discharge from the Ichetucknee springs, the chemical composition of their discharge water exhibits only minor variations through time. The response to recharge of the Ichetucknee springs may differ from other regions because of differences in the physical and chemical properties of the Floridan aquifer from previously studied karst aquifers. These properties could include age and types of rocks forming the aquifer, the amount of rainfall and its distribution through time and space, topography of the region, the depth to the water table, and the gradient of the potentiometric surface. The age and type of rocks in the aquifer would control their solubility and thus the extent and rate that reactions can alter the composition of the recharged water. The amount of rainfall, topography, depth to water table, and the potentiometric surface would be important because they control the volume of ground water that mixes with the recharged water and the residence time of the water in the subsurface. These results indicate that the formulation of a universal conceptual model of karst hydrogeology will require characterization of flow in several distinct aquifers, including the Floridan aquifer.

The relatively small changes in the composition of the discharged water through time are significant, even though the six springs in this study are tightly clustered along a short stretch of the river. These differences over such small distances reflect the extreme heterogeneity of porosity and permeability of karst aquifers. There are several important implications of this heterogeneity. Observations of seasonal and short term variations in composition at only one spring, or from one area, is unlikely to characterize the entire aquifer or even a single ground water basin that is drained by that spring. Complete characterization will likely require sampling several springs within a well-defined basin. The heterogeneity also implies that numerical models of physical flow and chemical reactions within basins will be difficult to achieve without conceptual constraints obtained through detailed observations of the composition, flow rates, and discharge volumes of the water.

Another important implication of this, and earlier work, has to do with water quality monitoring and regulation of karst aquifers, where porosity and permeability are heterogeneous. Small scale heterogeneity causes rapid changes in water composition and complex mixing of surface and ground water in the aquifer, as shown by the decoupling of NO_3, Cl, SO_4, Ca, and Mg concentrations seasonally at all of the Ichetucknee springs, and at Mill Pond Spring after the April storm. These rapid changes indicate that monitoring of water quality for contaminants will require multiple, and occasionally high frequency, sampling. The lack of systematic variations between rainy and dry seasons, but extreme variations after large storms (most clearly seen in the NO_3 concentrations), indicates that useful monitoring will require sampling more frequently during different seasons.

ACKNOWLEDGMENTS

Many people helped with this project in many different ways: Azell Nail at the Ichetucknee Springs State Park allowed free and unlimited access to the springs; Jim Stevenson at the Florida Department of Environmental Protection organized numerous interesting discussions through the Ichetucknee Springs Water Quality Working Group; Donald Monroe at the Suwannee River Water Management District provided the rainfall data; the Agronomy Department at the University of Florida provided air temperature data; Wes Skiles and Pete Butt of Karst Environmental Services organized the dye tracing study and provided us with preliminary results; and Brenda Eyre, Séverine Février, Brian Haley, Jennifer Miller, Bricky Way, and Benjamin Gordon provided field assistance. We are grateful to them all. Brian Katz provided numerous useful suggestions on an early version of the manuscript. The manuscript was also improved by thorough and helpful reviews by Eric Peterson and two anonymous reviewers. The research was supported by grants from the Department of Sponsored Research at the University of Florida and Sigma Xi, The Scientific Research Society.

REFERENCES

Andrews, W.J. 1994. Nitrate in ground water and spring water near four dairy farms in north Florida, 1990-1993: U.S. Geological Survey Water Resources Investigations Report 94-4162, United States Geological Survey, 63 p.

Beck, B.F. 1995. *Karst Geohazards, Engineering and Environmental Problems in Karst Terrane.* Rotterdam: A.A. Balkema, 579 p.

Boyer, D.G. & Pasquarell, G.C. 1995. Nitrate concentrations in karst springs in an extensively grazed area. *Water Resources Bulletin,* 31: 729-736.

Boyer, D.G. & Pasquarell, G.C. 1996. Agricultural land use effects on nitrate concentrations in a mature karst aquifer. *Water Resources Bulletin,* 32: 565-573.

Brooks, H.K. 1982. Physiographic divisions of the state of Florida: Gainesville, Florida, Cooperative Extension Service, Institute of Food and Agriculture Sciences, University of Florida, scale 1:500,000, 1 sheet.

Chapelle, F.H. & Knobel, L.L. 1986. Stable carbon isotopes of HCO_3 in the Aquia Aquifer, Maryland: Evidence for an isotopically heavy source of CO_2. *Ground Water,* 24: 592-599.

Chapelle, F.H., Morris, J.T., McMahon, P.B. & Zelibor, J.J.L. 1988. Bacterial metabolism and the $\delta^{13}C$ composition of ground water, Floridan aquifer system, South Carolina. *Geology,* 16: 117-121.

Dreiss, S.J. 1983. Linear unit-response functions as indicators of recharge areas for large karst springs. *Journal of Hydrology,* 61: 31-44.

Dreiss, S.J. 1989a. Regional scale transport in a karst aquifer 2. Linear systems and time moment analysis. *Water Resources Research,* 25: 126-134.

Dreiss, S.J. 1989b. Regional scale transport in a karst aquifer 1. Component separation of spring flow hydrographs. *Water Resources Research,* 25: 117-125.

Drever, J.I. 1988. *The Geochemistry of Natural Waters,* 2nd edition: Englewood Cliffs, New Jersey: Prentice Hall, 437 p.

Felton, G.K. & Currens, J.C. 1994. Peak flow rate and recession-curve characteristics of a karst spring in the Inner Bluegrass, central Kentucky. *Journal of Hydrology,* 162: 99-118.

Ford, D.C. & Williams, P.W. 1989. *Karst Geomorphology and Hydrology:* Winchester, Mass.: Unwin Hyman, Ltd., 601 p.

Gieskes, J.M., Gamo, T. & Brumsack, H. 1992. Chemical methods for interstitial water analyses on JOIDES Resolution: College Station, TX, Ocean Drilling Program, unpaginated.

Gordon, S.L. 1998. Surface and ground water mixing in an unconfined karst aquifer, Ichetucknee River ground water basin, Florida (M.S. thesis): Gainesville, University of Florida, 121 p.

Greene, E.A. 1997. Tracing recharge from sinking streams over spatial dimensions of kilometers in a karst aquifer. *Ground Water,* 35: 898-904.

Groszos, M., Ceryal, R., Allison, D., Cooper, R., Weinberg, M., Macesich, M., Enright, M.M. & Rupert, F. 1992. *Carbonate units of the intermediate aquifer system in the Suwannee River Water Management District,* Florida: Florida Geological Survey, 22 p.

Günay, G. & Johnson, A.I. 1986. *Karst Water Resources.* International Association of Hydrological Sciences, Ankara, Turkey, IAHS, 642 p.

Hess, J.W. & White, W.B. 1988. Storm response of the karstic carbonate aquifer of south-central Kentucky. *Journal of Hydrology,* 99: 235-252.

Hirth, D.K. 1995. Hydrogeochemical characterization of the Ichetucknee River groundwater basin using multiple tracers and computer modeling near Lake City, Florida (M.S. thesis): Gainesville: University of Florida, 115 p.

Hunn, J.D. & Slack, L.J. 1983. Water resources of the Santa Fe River Basin, Florida: U.S. Geological Survey Water Resources Investigations Report 83-4075, 105 p.

Iqbal, M.Z. & Krothe, N.C. 1995. Infiltration mechanisms related to agricultural waste transport through the soil mantle to karst aquifers of southern Indiana, USA. *Journal of Hydrology,* 164: 171-192.

Jones, I.C., Vacher, H.L. & Budd, D.A. 1993. Transport of calcium, magnesium and SO_4 in the Floridan aquifer, west-central Florida: Implications to cementation rates. *Journal of Hydrology,* 143: 455-480.

Katz, B.G. 1992. Hydrochemistry of the Upper Floridan Aquifer in Florida: U.S. Geological Survey Water Resources Investigations Report 91-4196, 37 p.

Katz, B.G. & DeHan, R.S. 1996. The Suwannee River basin pilot study: Issues for watershed management in Florida: U.S. Geological Survey Fact Sheet FS-080-96, 3 p.

Katz, B.G. & Hornsby, H.D. 1998. A preliminary assessment of sources of nitrate in springwaters, Suwannee River Basin, Florida: U.S. Geological Survey Open File Report 98-69, 18 p.

Katz, B.G., Lee, T.M., Plummer, L.N. & Busenberg, E. 1995a. Chemical evolution of groundwater near a sinkhole lake, northern Florida: 1. Flow patterns, age of groundwater, and influence of lakewater leakage. *Water Resources Research*, 31: 1549-1564.

Katz, B.G., Plummer, L.N., Busenberg, E., Revesz, K.M., Jones, B.F. & Lee, T.M. 1995b. Chemical evolution of groundwater near a sinkhole lake, northern Florida: 2. Chemical patterns, mass-transfer modeling, and rates of chemical reactions. *Water Resources Research*, 31: 1565-1584.

Katz, B.G., Coplen, T.B., Bullen, T.D. & Davis, J.H. 1997a. Use of chemical and isotopic tracers to characterize the interactions between ground water and surface water in mantled karst. *Ground Water*, 35: 1014-1027.

Katz, B.G., DeHan, R.S., Hirten, J.J. & Catches, J.S. 1997b. Interactions between ground water and surface water in the Suwannee River Basin, Florida. *Journal of the American Water Resources Association*, 33: 1-18.

Kincaid, T.R. 1997. Ground water-Surface water exchange in the unconfined karstified Floridan aquifer, in Günay G. & Johnson A.I. (eds) *Karst Waters and Environmental Impacts*. Rotterdam: A.A. Balkema, p. 405-412.

Lakey, B. & Krothe, N.C. 1996. Stable isotopic variation of storm discharge from a perennial karst spring, Indiana. *Water Resources Research*, 32: 721-731.

LaMoreaux, P.E., Powell, W.J. & LeGrand, H.E. 1997. Environmental and legal aspects of karst areas. *Environmental Geology*, 29: 23-36.

Pasquarell, G.C. & Boyer, D.G. 1996. Herbicides in karst groundwater in southeast West Virginia. *Journal of Environmental Quality*, 25: 755-765.

Padilla, A., Pulido-Bosch, A. & Mangin, A. 1994. Relative importance of baseflow and quickflow from hydrographs of karst springs. *Ground Water*, 32: 267-277.

Pitty, A.F. 1968. Calcium carbonate content of water in relation to flow-through time. *Nature*, 217: 939-940.

Plummer, L.N. 1977. Defining reactions and mass transfer in part of the Floridan Aquifer. *Water Resources Research*, 13: 801-812.

Pucci, A.A., Ehlke, T.A. & Owens, J.P. 1992. Confining unit effects on water quality in the New Jersey Plain. *Ground Water*, 30: 415-427.

Puri, H.S. & Vernon, R.O. 1964. *Summary of the Geology of Florida and a Guidebook to the Classic Exposures*. Florida Geological Survey Special Publication 5, 312 p.

Rosenau, J.C., Faulkner, G.L., Hendry, J.C.W. & Hull, R.W. 1977. Springs of Florida: Florida Department of Natural Resources, Tallahassee, Florida, *Bulletin* no. 31, 461 p.

Ryder, P.D. 1985. Hydrology of the Floridan Aquifer System in West-Central Florida: U.S. Geological Survey Professional Paper 1403-F, 63 p.

Sacks, L.A., Herman, J.S. & Kauffman, S.J. 1995. Controls on high sulfate concentrations in the upper Floridan aquifer in southwest Florida. *Water Resources Research*, 31: 2541-2551.

Saunders, J.A. & Toran, L.E. 1994. Evidence for dedolomitization and mixing in paleozoic carbonates near Oak Ridge, Tennessee. *Ground Water*, 32: 207-214.

Scott, T.M. 1988. The lithostratigraphy of the Hawthorn Group (Miocene) of Florida. *Florida Geological Survey Bulletin* no. 59, 147 p.

Scott, T.M. 1992. A Geological Overview of Florida: Florida Geological Survey Open File Report no. 50, 78 p.

Shuster, E.T. & White, W.B. 1971. Seasonal fluctuations in the chemistry of limestone springs: A possible means for characterizing carbonate aquifers. *Journal of Hydrology*, 14: 93-128.

Shuster, E.T. & White, W.B. 1972. Source areas and climatic effects in carbonate groundwaters determined by saturation indices and carbon dioxide pressures. *Water Resources Research*, 8: 1067-1073.

Smart, C.C. & Ford, D.C. 1986. Structure and function of a conduit aquifer. *Canadian Journal of Earth Science*, 23: 919-929.

Sprinkle, C.L. 1989. Geochemistry of the Floridan Aquifer system in Florida and in parts of Georgia, South Carolina, and Alabama: U.S. Geological Survey Open File Report, 105 p.

Stringfield, V.T. 1966. Artesian water in tertiary limestone in the southeastern states. *Geological Society of America Special Paper* no. 93, 226 p.

Ternan, J.L. 1972. Comments on the use of a calcium hardness variability index in the study of carbonate aquifers with reference to the central Pennines, England. *Journal of Hydrology*, 16: 317-321.

Thornthwaite, C.W. 1948. An approach toward a rational classification of climate. *Geographical Review*, 38: 55-94.

Tibbals, C.H. 1990. Hydrology of the Floridan Aquifer System in East-Central Florida. *U.S. Geological Survey Professional Paper* 1403-E, 98 p.

Upchurch, S.B. 1992. Quality of water in Florida's aquifer systems, in Maddox, G.L., Lloyd, J.M., Scott, T.M., Upchurch, S.B. & Copeland, R. (eds) Florida's Ground Water Quality Monitoring Program, Background Hydrogeochemistry: Tallahassee, Florida, *Florida Geological Survey*, p. 12-51.

Upchurch, S.B. & Lawrence, F.W. 1984. Impact of ground-water chemistry on sinkhole development along a retreating scarp, in Beck, B.F. (ed.) *Sinkholes: Their Geology, Engineering and Environmental Impact.* Rotterdam: A.A. Balkema, p. 189-195.

White, W.A. 1970. The geomorphology of the Florida Peninsula. *Florida Geological Survey Geological Bulletin* no. 51., 164 p.

White, W.B. 1988. *Geomorphology and Hydrology of Karst Terrains.* New York: Oxford University Press.

White, W.B., Culver, D.C., Herman, J.S., Kane, T.C. & Mylroie, J.E. 1995. Karst lands. *American Scientist*, 83: 450-459.

Wicks, C.M. & Herman, J.S. 1994. The effect of a confining unit on the geochemical evolution of ground water in the Upper Floridan aquifer system. *Journal of Hydrology*, 153: 139-155.

Wicks, C.M., Herman, J.S., Randazzo, A.F. & Gee, J.L. 1995. Water-rock interactions in a modern coastal mixing zone. *Geological Society of America Bulletin*, 107: 1023-1032.

Wigley, T.M.L. & Plummer, L.N. 1976. Mixing of carbonate waters. *Geochimica et Cosmochima Acta*, 40: 989-995.

Estimating porosity and permeability in a karstic aquifer using core plugs, well tests, and outcrop measurements

ROBERT E. MACE & SUSAN D. HOVORKA
Bureau of Economic Geology, The University of Texas at Austin, Austin, Texas, USA

ABSTRACT: The statistical and spatial distribution of matrix and conduit/fracture porosity and permeability is important for addressing water resource and contaminant transport issues in karstic aquifers. We used quantitative resource assessment methods, well performance tests, geostatistics, and fractal scaling to develop an integrated understanding of porosity and permeability in the Edwards Group in South-Central Texas. Using porosity and permeability determined with plugs collected from core and outcrop, wireline log interpretation, and three-dimensional geocellular modeling software used in hydrocarbon resource quantification, we estimated 1. the three-dimensional distribution of matrix porosity and permeability in the aquifer, 2. the porosity due to dissolution features, and 3. the volumetrics of the aquifer. Matrix porosity measured in core plugs ranged from 0.7 to 49.8% with a mean of 14.4%. Vuggy porosity (dissolution features) estimated from core plugs is about 2%. Hydraulic conductivity of the matrix determined from core plugs ranged from 8.4×10^{-6} to 8.3 m d^{-1} with a geometric mean of 1.3×10^{-3} m d^{-1}. Estimated total volume of water-filled pore space in the Edwards aquifer is 2.13×10^{11} m^3. Transmissivity of the aquifer due to matrix, fractures, and dissolution features was estimated using more than 1000 well tests and was found to be log-normally distributed, vary over eight orders of magnitude, have a geometric mean of 540 m^2 d^{-1}, and be spatially correlated. We estimated the porosity of fractures and dissolution features at roadcut exposures using scanlines, image analysis, and fractal scaling. We found that secondary porosity in the outcrops due to fractures is 0.2 to 0.7% and secondary porosity due to dissolution features is 2.5 to 10%.

1 INTRODUCTION

As water resources become more heavily utilized by a growing population, more detailed investigations of aquifers need to be undertaken to better understand ground-water flow and quantify the extent and status of ground-water resources. With greater interest in ground-water resources, techniques previously reserved for enhancing recovery from hydrocarbon reservoirs can now be applied to aquifers to better understand and quantify porosity and permeability. Investigations of karstic aquifers are particularly challenging because complex networks of fractures and dissolution features can locally and regionally control ground-water flow.

The Edwards aquifer of south-central Texas is a geologically complex karstic water resource that is the sole source of water for the city of San Antonio and the main water resource for agriculture and industry in surrounding counties. Discharge from the aquifer feeds springs that are recreational centers, habitat for endangered species, and important sources of water to rivers that are used for recreation, agriculture, and industry. During drought periods, ground-water production exceeds recharge resulting in lower water levels and substantial decreases in springflow. Maps of the geologically controlled distribution of porosity and permeability are needed to better understand and quantify volumetrics, flow, and transport in the aquifer and are required to develop more detailed and accurate numerical models.

Because karstic aquifers are so complex, a single technique cannot fully characterize the porosity and permeability. Furthermore, the application of many standard techniques is difficult and limited in karstic aquifers. However, using multiple techniques (with an appreciation of their strengths and limits) significantly helps to develop a better overall understanding of how porosity and permeability are distributed in the aquifer. Multiple techniques provide insights on aquifer architecture and ground-water flow that are useful for ground-water management and for addressing ongoing legal debates on ground-water ownership and spring discharge.

This paper is meant as an integration and comparison of the diverse techniques we employed to characterize porosity and permeability in a karstic aquifer. We have previously published papers and reports on the results of the study and the application of some of the techniques to the Edwards aquifer (Hovorka et al., 1996, 1998; Mace, 1995, 1997). However, this paper is the first to focus on comparing and contrasting results of multiple measurement techniques for quantifying porosity and permeability in a karstic aquifer.

2 STUDY AREA

The Balcones Fault Zone, or San Antonio, segment of the Edwards aquifer is a hydrostratigraphic unit that includes limestones of the Edwards Group (120 to 180 m thick with greater thickness further downdip) and the Georgetown Formation (< 18 m thick) and extends over an area of about 7800 km^2 in south-central Texas (Fig. 1) that includes Bexar, Comal, Hays, Kinney, Medina, and Uvalde Counties (Maclay & Small, 1986; Maclay, 1995). The aquifer is bounded to the north-east by the Kyle ground-water divide, the west by the Brackettville ground-water divide, the north and west by the outcrop, and the south and east by a fresh/saline-water interface commonly known as the bad-water line (Schultz, 1993). The Edwards aquifer is confined where overlain by younger, low-permeability rocks and bounded on the bottom by low permeability carbonates.

Three major depositional belts compose the Edwards Group: San Marcos Platform (Rose, 1972), Devils River trend (Rose, 1972), and Maverick Basin (Smith, 1964) (Fig. 1). The San Marcos platform consists of the Kainer and Person Formations (Rose, 1972) and is characterized by high-frequency depositional cycles that include subtidal wackestones and packstones, grain-dominated packstones, grainstones, subtidal gypsum (replaced by calcite), and dolomitic tidal-flat wackestones and grainstones (Hovorka et al., 1996). The Devils River Formation (Rose, 1972) was deposited on the platform margin and is lithologically similar to the Kainer and Person Formations (Hovorka et al., 1996). The Maverick Basin consists of the West Nueces (predominately low-porosity sub-

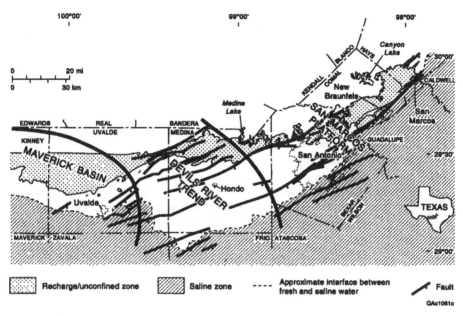

Figure 1. Location of the Balcones Fault Zone aquifer in south-central Texas including the facies (after Hovorka et al., 1995), structural (after Collins & Hovorka, 1997), and hydrologic setting (after Hovorka et al., 1996). The confined zone is between the recharge and saline zones.

tidal wackestones and packstones), McKnight (highly cyclic subtidal dark, organic-rich, argillaceous limestones), and Salmon Peak (massive, burrowed, miliolid packstone) Formations and were deposited in subtidal, slightly deeper water environments (Hovorka et al., 1996).

The Edwards aquifer lies in and is intimately associated with the Balcones Fault Zone, a regional feature of normal faults due to gulfward extension, flexure, and tilting along the perimeter of the Gulf of Mexico that extends from near Del Rio, through San Antonio, up through New Braunfels, Austin, and Waco, and towards Dallas (Murray, 1961; Maclay & Small, 1983; Ewing, 1991). Normal faults of the Balcones Fault Zone (Fig. 1) have throws as great as 250 m and are the main structural control on the Edwards aquifer (Maclay & Small, 1983; Collins & Hovorka, 1997).

The Edwards aquifer can be divided into three main hydrogeologic settings: the recharge zone, the confined zone, and the saline zone (Fig. 1). The Edwards aquifer outcrop belt approximately defines the main recharge zone and unconfined part of the aquifer. The confined zone is approximately defined where the Edwards aquifer is overlain by confining units and where total dissolved solids is less than 1000 ppm. The saline zone is confined and has total dissolved solids greater than 1000 ppm (Groschen & Buszka, 1997).

Fresh-water circulation and development of the modern aquifer followed uplift along extensional faults of the Balcones Fault system. Integration of the aquifer as a result of erosion, valley incision, and karst piracy (Woodruff & Abbott, 1986) probably allowed deeper penetration of fresh water into the Edwards Group, resulting in a well-developed network of solutionally enlarged fractures (Maclay & Small, 1986). The Edwards aquifer is vital to the regional economy (San Antonio is the largest U.S. city solely dependent on ground water), supports habitats for several endangered species in its springs, and is threatened by continued urbanization and development (Sharp & Banner, 1997).

3 APPROACH

Our approach involved quantifying porosity and permeability using several different methods and comparing and contrasting results. The different techniques include: 1. measuring porosity and permeability using core and outcrop plug samples, 2. estimating porosity from geophysical logs, 3. estimating matrix permeability from porosity, 4. interpolating matrix porosity and permeability and estimating aquifer volumetrics using a three-dimensional geocellular model, 5. estimating well-scale permeability, 6. geostatistically analyzing transmissivity, and 7. investigating scale invariance of fracture apertures and cross-sectional area and hydraulic diameter of dissolution features and using the scale invariance to estimate fracture and conduit porosity and permeability in the aquifer.

3.1 *Core and outcrop sample analysis*

We collected 506 horizontally-oriented core plugs (2.54 cm in diameter) from eight cores and five outcrops representative of the Edwards aquifer. Core plugs were collected at approximately 30 cm intervals over eight 6 to 15 m intervals selected for study and were analyzed for porosity and permeability using standard oil field techniques.

3.2 *Geophysical log interpretation and porosity estimation*

Neutron and resistivity logs of water, oil, and gas wells through the Edwards aquifer were obtained from USGS and TWDB studies (Sieh, 1975; Maclay & Small, 1976; Maclay et al., 1981), commercial logs from the Surface Casing Division of the Texas Water Commission, and logs from the Edwards Underground Water District. Logs that contained usable porosity information were digitized at 30 cm increments.

Neutron log response was empirically calibrated to porosity by comparing it to running averages of porosity measured in core plugs (Hovorka et al., 1996). Caliper-corrected neutron-log response was cross plotted against plug-derived porosity to produce a relationship between measured porosity and log response which was used to estimate porosity for other neutron logs (Fig. 2). This relationship estimates porosity to within approximately 5 percentage points.

Porosity of the matrix, ϕ_m (in percent), was calculated from resistivity logs using the standard equation (Archie, 1942)

$$\phi_m = 100 \sqrt{\frac{R_{wa}}{R_t}} \tag{1}$$

where R_{wa} is the resistivity of water and R_t is the calibrated resistivity value and assuming a cementation factor of 2 (Schultz, 1992). R_{wa} was determined from

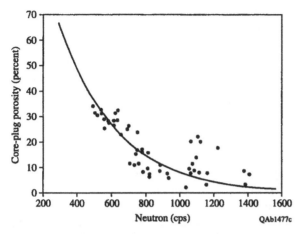

Figure 2. Neutron-log response in cps versus core-plug porosity (dots) and calculated porosity using caliper and depth correction (line) (from Hovorka et al., 1996, p. 9).

$$R_{wa} = \frac{10,000}{C} \qquad (2)$$

where C is the specific conductance of water in the formation determined from a revised map of specific conductance (Hovorka et al., 1996). Various correction techniques were used to correct poorly-scaled logs and non-linear freshwater responses (Hovorka et al., 1996). We interpreted porosity values larger than 50% to indicate that the well bore penetrated a macroscopic vug or cave.

3.3 *Estimation of matrix permeability*

Matrix permeability was estimated in the aquifer by developing relationships between permeability and porosity from core plugs for different rock types and using these relationships to estimate permeability from porosity estimated from neutron-log response (Hovorka et al., 1998). Grouping the samples by depositional and aquifer setting and then considering the mineralogy and petrographic characteristics reduced the scatter and provided usable relationships for calculating matrix permeability. Porosity-permeability relationships were derived for different geologic and hydrogeologic settings (Fig. 3). These relationships generally estimate matrix permeability to within an order of magnitude. The relationships were used to estimate matrix permeability for each 30-cm interval in 127 boreholes where porosity had been estimated from neutron or resistivity logs. Permeability estimated from porosity which had been estimated from neutron-log response should be within one and a half orders of magnitude of its actual value. A porosity cutoff of 35% was used to eliminate samples dominated by fractures and dissolution features from the matrix permeability calculation.

3.4 *Three-dimensional geocellular modeling*

Interwell porosity and permeability interpolations were created using a three-dimensional model built with Stratamodel© Stratigraphic Geocellular Modeling (SGM) software (Hovorka et al., 1996, 1998). The physical limits of the aquifer (top, bottom, bad-water line, water-level surface in the unconfined zone, and ground-water divides) were input based on controlled structural data (Collins & Hovorka, 1997). The aquifer was divided into 196 layers within stratigraphic horizons. Interpolation of porosity and permeability input for each 30-cm interval between wells was done using least-squares regression for each cell in the model and imaged in map view, cross section, and block diagram. Total water-filled void volume of the aquifer was calculated by summation.

3.5 *Estimation of well-scale permeability*

We compiled well-test data, including pumping, step-drawdown, and specific-capacity tests, from open-file records at the Texas Water Development Board and the Texas Natural Resource Information System and from research and technical reports (Myers, 1969; Klemt et al., 1979; Marquardt & Elder, 1979; Maclay et al., 1980; Guyton et al., 1986; Alexander, 1990; Poteet et al., 1992). A total of 1083 specific-capacity and transmissivity tests were compiled with a range of aquifer penetration. Of these tests, 71 were time-drawdown tests, 32 were step-drawdown tests, and 980 were specific-capacity tests. For the pumping tests, transmissivity was estimated using the Theis (1935) or the Cooper & Jacob (1946) method on drawdown data and the Theis (1935) recovery method on recovery data.

Transmissivity was estimated from the specific capacity tests using an empirical relationship between transmissivity, T [m^2 d^{-1}], and specific capacity, S_C [m^2 d^{-1}]

$$T = 0.96(S_C)^{.08} \tag{3}$$

This empirical relationship was derived by fitting a line to a log-log plot of measured transmissivity and specific capacity (Mace, 1997). This relationship has a 95% prediction interval of 1.6 orders of magnitude (Fig. 4), which means that the range of probable transmissivities for a given measured specific capacity is more than one and a half orders

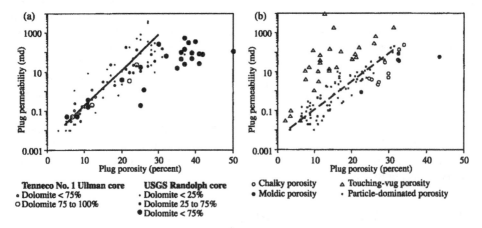

Figure 3. Representative porosity-permeability relationships for the Edwards aquifer (from Hovorka et al., 1998). Plots also show the relationship between porosity and permeability and dolomite, pore type, and pore structure.

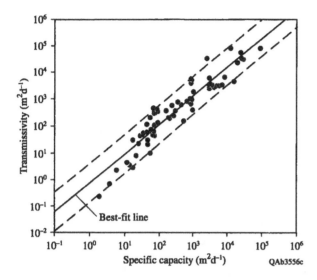

Figure 4. Relationship between specific capacity and transmissivity for the Edwards aquifer (from Mace, 1997) showing the data (solid points), best fit line (solid line), and the 95% prediction intervals (dashed lines).

of magnitude. Though this prediction interval is large, the relationship and the range of specific capacity in the Edwards aquifer spans over five orders of magnitude resulting in useful estimates of transmissivity (Mace, 1997). Hydraulic conductivity was determined by dividing the estimated transmissivity by the well interval open to the Edwards aquifer. For fully penetrating wells, this was the thickness of the aquifer.

While compiling well-test data we noticed that there were specific capacity tests with no measurable drawdown in the same area as tests with comparable pumping rates and measurable drawdown. We decided that not including these tests would bias transmissivity to lower values. To include these zero-drawdown tests, we assumed that most operators could measure a 0.3 meter change in water level (estimated by inspecting files and interviewing well drillers) and calculated a minimum specific capacity value based on this drawdown.

3.6 *Geostatistical analysis of transmissivity*

The spatial correlation of transmissivity in the aquifer was investigated by generating semivariograms, which is a plot of the semivariance versus distance (Clark, 1979). Based on the structure of the experimental semivariogram, two theoretical semivariograms were used to describe the semivariance of transmissivity: a fractal Brownian motion (fBm) semivariogram and a Gaussian semivariogram. The fBm semivariogram was used to model semivariance for both small and large scales and was used to krige regionally. The Gaussian semivariogram was used to better model small-scale semivariance for local kriging.

The fBm semivariogram is described by

$$\gamma(r) = \begin{cases} N + (S-N)\left(\dfrac{r}{R}\right)^{2H} & r < R \\ S & r \geq R \end{cases} \qquad (4)$$

where $\gamma(r)$ is the semivariance for separation distance r, N is the nugget, S is the sill, R is the range, and H is the fractal co-dimension.

The Gaussian semivariogram (Isaaks & Srivastava, 1989, p. 375, adjusted here to include a nugget) is described by

$$\gamma(r) = N + (S - N)\left(1 - e^{-3\left(r^2/R^2\right)}\right) \tag{5}$$

We kriged transmissivity using the fBm theoretical semivariogram at a grid size of 3,500 m (half the range of the small-scale correlation). Kriging was performed on the mean transmissivity of the wells contained within the grid cell.

3.7 *Analysis of fractal scaling*

We used feeler gauges and image analysis to characterize and interpret hundreds of fracture apertures and the cross-sectional area of thousands of dissolution features in eight exposures in a variety of structural and stratigraphic settings (Hovorka et al., 1998). Using automotive feeler gauges with a range of 0.038 to 0.889 mm and a 30-cm metal ruler for larger features, we measured fracture aperture at three of the exposures along 20- to 30-m transects. We took numerous photographs along the exposures, careful to maintain a consistent scale and minimize parallax effects. After assembling the photographs into a photomosaic (e.g., Fig. 5a), we traced the limits of the exposure and the size and location of dissolution features onto Mylar sheets to eliminate shadow, vegetation, and surface roughness effects (Fig. 5b) and scanned the sheets to create binary (black and white) digital image files for image analysis. We calculated the area of the dissolution features using the public domain program NIH image (Rasband, 1994).

Because the measurements were made at land surface, fractures and dissolution features have undoubtedly been effected by unloading, weathering, solution, and mechanical disturbance. Fractures measured under these conditions will likely have greater apertures than those in the subsurface. Although modern processes have influenced karst development differently in the confined, unconfined, and unsaturated zones of the aquifer, we believe that much of the fabric observed in outcrop is inherited from earlier conditions and is representative of the subsurface.

We tested for scale invariance (fractal behavior) of fracture aperture and cross-sectional area of dissolution features using frequency-size distribution plots (Harris et al., 1991; Gillespie et al., 1993; Turcotte, 1997). Frequency-size distribution plots are made by plotting the cumulative number of features, N, greater than the feature attribute, A, against the value of the feature attribute (Fig. 6). If the distribution is scale invariant, the data will plot as a straight line, with

$$N = R\omega A^{-D}, \tag{6}$$

where R is the size of the sampling domain depending on the topology (length, R_l, for linear sampling; area, R_a, for two-dimensional sampling; and volume, R_v, for three-dimensional sampling), and ω and D are empirical fitting parameters, where D is the fractal dimension or power-law exponent.

Linearity on a frequency-size distribution plot suggests that the data are scale invariant over the range of observed values and can be described with a fractal (Fig. 6). Sampling bias can affect the collection of fracture data from exposures (Barton & Zoback, 1992), inducing artifacts in plots generated by the frequency-size distribution technique.

The summation of discrete measurements normalized to size of the investigated area (direct summation) can be used to determine aggregate properties such as porosity and permeability. If distributions are scale invariant, it is more accurate to determine aggregate properties recognizing the fractal scaling of the individual attributes (Marrett, 1996) so that under-represented and censored features can be fully included in aggregate calculations. The Riemann zeta function (Apostal, 1957), $\zeta(x)$, is a general expression of aggregate properties in terms of individual attributes following fractal scaling (Marrett, 1996) and is approximated by the sum of its first three terms and the first two terms of the Euler-Maclauren summation formula (Dahlquist & Björck, 1974),

$$\zeta(x) \cong 1 + 2^{-x} + 3^{-x} + \frac{x+7}{2(x-1)} 4^{-x},$$ (7)

where x is the argument.

Marrett (1996) showed that the aggregate fracture porosity, ϕ_f', sampled along a transect perpendicular to the power-law fractures is

(a)

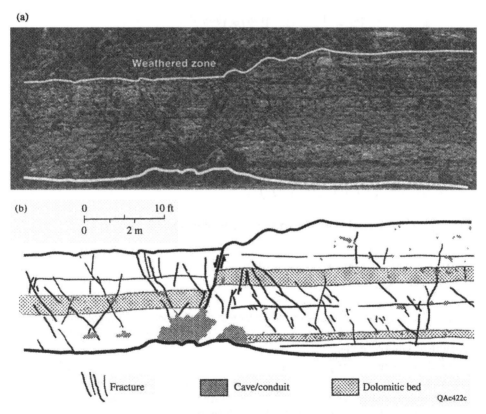

(b)

Figure 5. Example of a (a) photomosaic and (b) interpreted distribution of conduit porosity from part of the Lake Medina exposure.

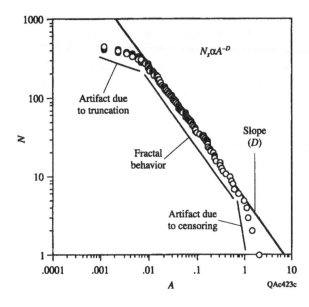

Figure 6. Example of a frequency-size distribution plot showing the effect of truncation and censoring bias on the data. N is the rank of the cross-sectional area of dissolution feature, A.

$$\phi'_f = \zeta\left(\frac{1}{D_f}\right)\frac{e_1}{R_l} = \zeta\left(\frac{1}{D_f}\right)\left(\omega_f\right)^{1/D_f}\left(R_l\right)^{\left(1/D_f - 1\right)}, \tag{8}$$

where e_1 is the largest aperture predicted by fractal scaling (the intercept of the fractal relationship with the x-axis when N equals 1 in Fig. 6), ω_f is a fitting parameter (see Equation 6) for fracture aperture, and D_f is the fractal dimension for fracture aperture. If we assume that the cubic law applies, then the aggregate permeability for the fracture set, k'_f, is estimated from (Marrett, 1996)

$$k'_f = \frac{1}{12}\zeta\left(\frac{3}{D_f}\right)\frac{e_1^3}{R_l} = \frac{1}{12}\zeta\left(\frac{3}{D_f}\right)\left(\omega_f\right)^{3/D_f}\left(R_l\right)^{\left(3/D_f - 1\right)}, \tag{9}$$

Similarly, the aggregate two-dimensional porosity of dissolution features is

$$\phi'_c = \zeta\left(\frac{1}{D_c}\right)\frac{A_1}{R_a} = \zeta\left(\frac{1}{D_c}\right)\left(\omega_c\right)^{1/D_c}\left(R_a\right)^{\left(\left(1/D_c\right) - 1\right)}, \tag{10}$$

where A_1 is the cross-sectional area of the largest dissolution feature predicted by fractal scaling, D_C is the fractal dimension, and ω_C is a fitting parameter for the cross-sectional area of dissolution features.

The above estimates of permeability assume laminar flow and smooth walls for the fracture or conduit. Furthermore, these estimates assume that each fracture or conduit in the distribution contributes to the total permeability and that each feature is infinite and linear in extent. These assumptions likely result in predicted permeabilities much higher than directly observed in the aquifer.

4 RESULTS

Results of the different techniques are summarized below with a discussion that compares, contrasts, and integrates the estimates of porosity and permeability.

4.1 *Porosity*

Matrix porosity measured in core plugs ranges from 0.7 to 49.8% with a mean of 14.4%. Average porosity through the thickness of the aquifer measured using logs generally shows lateral gradational changes from a minimum of 16% to a maximum of 28% with an average of 18%. Vuggy porosity (> 50% porosity on porosity logs) is about 2%. The variation of porosity distribution is due to high-frequency cyclicity, diagenesis history, and the deposition of the formations (Hovorka et al., 1996). Estimated total volume of water-filled pore space in the Edwards aquifer is 2.13×10^{11} m^3. The unconfined zone stores 1.87×10^{10} m^3, and the confined zone stores 1.94×10^{11} m^3. The amount of water in the unconfined zone between the historic high and low water levels is 8.63×109 m^3. On average, about 8.64×10^8 m^3 recharges the aquifer each year (Brown et al., 1992).

The distribution of fracture aperture and the cross-sectional area of dissolution features in the investigated exposures follow power laws with fractal dimensions of 0.85 to 0.93 for aperture and 0.59 to 0.95 for conduit cross-sectional area. Assuming fractal scaling, we estimated secondary porosity in the outcrops due to fractures to be 0.2 to 0.7% and due to dissolution features to be 2.5 to 10%. When calculated by direct summation, secondary porosity of the outcrops due to vertically oriented fractures and dissolution features ranges between 0.07 and 0.3% and between 1.1 to 5.7%, respectively. For the investigated exposures, geometric mean fracture aperture or dissolution feature cross-sectional area is not correlated to structural or stratigraphic setting, although the comparison may be limited by overgeneralizing structural and stratigraphic setting.

4.2 *Permeability*

Hydraulic conductivity of the matrix determined from core plugs ranged from 8.4×10^{-6} to 8.3 m d^{-1} with a geometric mean of 1.3×10^{-3} m d^{-1}. Average hydraulic conductivity of

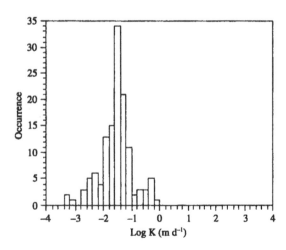

Figure 7. Histogram of geometrically-averaged matrix hydraulic conductivity for the Edwards aquifer.

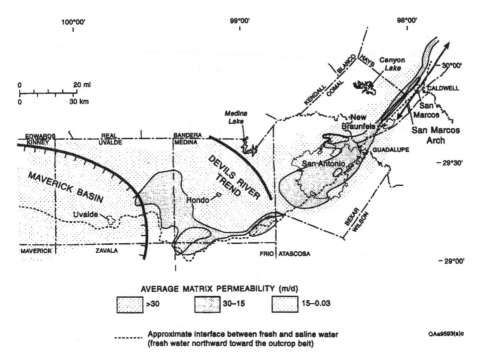

Figure 8. Distribution of vertically averaged matrix permeability (from Hovorka et al., 1998).

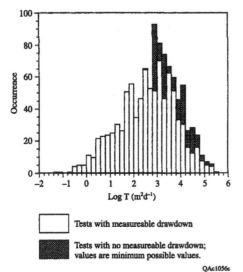

Figure 9. Histogram of transmissivity in the Edwards aquifer including values based on measured drawdowns and estimated minimum values for tests with no drawdowns.

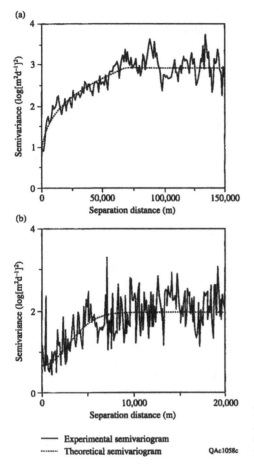

Figure 10. Experimental and theoretical semivariograms for transmissivity at (a) large and (b) small scales. Theoretical semivariogram for (a) is the fBm model and for (b) is the Gaussian model.

the matrix calculated for each logged well is approximately log normally distributed, extends over three and a half orders of magnitude, and has a geometric mean of 0.03 m d^{-1} (Fig. 7). Matrix permeability is enhanced near the bad water line (Fig. 8) probably as a result of increased dolomite dissolution due to the mixing of fresh and saline water (Deike, 1990).

Hydraulic conductivity for fractures and dissolution features determined from outcrops measurements using direct summation ranges from 140 to 12,000 m d^{-1} and from 4.8 × 10^6 to 1.5 × 10^8 m d^{-1}, respectively. Hydraulic conductivity estimated from fractal scaling ranges from 110 to 35,000 m d^{-1} for fractures and 1.3 × 10^7 to 3.8 × 10^8 m d^{-1} for dissolution features. These estimates are unrealistically high and are discussed in more detail in the next section.

Transmissivity is log-normally distributed and varies over eight orders of magnitude with a geometric mean of 540 m^2 d^{-1} (Fig. 9). Not including the zero-drawdown tests results in a mean transmissivity of 360 m^2 d^{-1}. Hydraulic conductivity determined from well tests has a geometric mean of 3.3 m d^{-1}, more than 100 times larger than that for the matrix.

Observed heterogeneity is related to tectonic fracturing along a normal fault system; preferential conduit enlargement in the subtidal dolomitic parts of stacked high-frequency cycles; regional platform-to-basin facies patterns; and aquifer development in a mixing zone of fresh and saline water. The influence of stratigraphy and geology on the distribution of porosity and permeability in the aquifer are further analyzed in Hovorka et al. (1996, 1998).

Semivariograms show that transmissivity is spatially correlated and that semivariance can be modeled for two scales: 5 km at the small scale and 75 km at the regional scale (Fig. 10). Different semivariogram structure at the smaller scale may be due to more similar aquifer attributes between major faults or within a facies type. Spatial correlation at the larger scale may be a function of the depositional belts or the regional flow path. The semivariograms have large nuggets (Fig. 10) which reflects apparent small-scale randomness possibly due to measurement error, the complex interaction of the variables determining specific capacity and transmissivity, and real variation in permeability.

An fBm semivariogram (Equation 4) was fit to the large-scale empirical semivariogram for transmissivity (Fig. 10a) where N, S, R, and H are 0.4 (log m^2 d^{-1})^2, 2.9 (log m^2 d^{-1})^2, 70,000 m, and 0.16, respectively. A Gaussian semivariogram (Equation 5) was fit to the small-scale empirical semivariogram for transmissivity (Fig. 10b) where N, S, and R are 0.6 (log m^2 d^{-1})^2, 1.9 (log m^2 d^{-1})^2, and 7000 m, respectively. Our previous analyses approximated the experimental semivariogram with a spherical semivariogram (Mace, 1995; Hovorka et al., 1998).

Smoothed, kriged transmissivity indicates higher transmissivity around San Antonio and through the middle of Medina and Uvalde Counties (Fig. 11). Uncertainty of kriging estimates are higher for Frio, Kinney, Edwards, Real, and the northern part of Uvalde Counties (Hovorka et al., 1998).

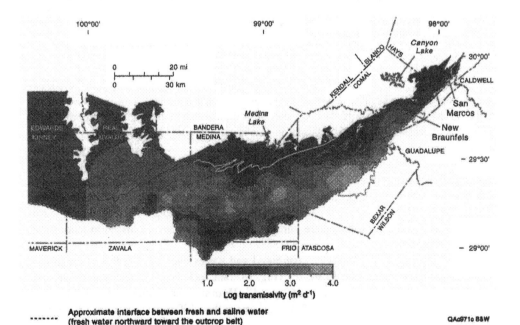

Figure 11. Smoothed kriged map of transmissivity (after Hovorka et al., 1998).

5 DISCUSSION

We used several different techniques to quantify porosity and permeability. Some of the techniques are interrelated. Log-based porosity was calibrated using empirical relationships of log-response to porosity measured in core-plugs. Log-based matrix permeability is derived from matrix permeability measured in plugs and porosity estimated from log response. Porosity and permeability estimates using fractal scaling are based on the direct measurements of fractures and dissolution features at the outcrop.

The different techniques offer differing results. Porosity estimated from geophysical logs has a larger mean (18%) and a smaller range (16 to 28%) than the mean and range of porosity measured in plugs (14.4%, 0.7 to 49.8%, respectively). The smaller range of log-estimated porosity is due to vertically averaging matrix porosity which removes the effects of high frequency vertical variability in bedded carbonate. Plug porosity is biased toward low values because of poor recovery of many of the highly porous friable or vuggy intervals and deliberate over-sampling of thin, low-porosity intervals for log calibration.

Fracture porosity estimated using fractal scaling of fractures measured at surface exposures (0.2 to 0.7%) were slightly larger than that determined by direct summation (0.07 to 0.3%). This is because porosity estimated with fractal scaling accounts for smaller and larger fractures not measured in the outcrop due to censoring and truncation. Similarly, porosity of dissolution features estimated using fractal scaling (2.5 to 10%) were slightly larger than that determined by direct summation (1.1 to 5.7%). Vuggy porosity (suggesting the presence of large karst-related dissolution features or honeycombed breccia) determined from log response is at the lower end of outcrop-based dissolution-feature porosity estimates (2%). Outcrop-based measurements are biased toward higher values due to unloading effects and surface weathering. The resolution of commercial logs is inadequate to exactly define dissolution features less than 30 cm and biases estimates toward smaller values. Geophysical logs with better resolution would result in better estimates of porosity due to fractures and dissolution features. Furthermore, log-based estimates are based on one-dimensional sampling of three dimensional features which further underestimates porosity.

Vertically averaged permeability estimated from logs has a larger mean (0.03 m d^{-1}) than the mean permeability measured in plugs (0.0013 m d^{-1}). Similar to porosity, plug porosity is biased toward low values because of undersampling highly porous friable or vuggy intervals and deliberate over-sampling of low-porosity intervals for log calibration. It is also very sensitive to the cutoff applied for removing vuggy porosity. Permeability based on well-testing is considerably higher (3.3 m d^{-1}) than vertically averaged permeability estimated from logs (0.03 m d^{-1}). This is because well-based values sample the entire aquifer interval including all fractures and conduits, whereas log-based values predominantly represent the matrix. The interpolated map of matrix permeability shows higher values along the bad-water line while the transmissivity map based on well tests has higher values further north. This difference is probably because there are fewer well tests near the bad-water line to fully characterize transmissivity in that area.

Permeability estimated from outcrop measurements of fractures and dissolution features are considerably higher than measured values. This is because the technique assumes that each fracture and dissolution feature have an infinite length with the same measured size over that length. In reality, many of the fractures and most of the dissolution features are not interconnected, which would lead to much lower estimates of permeability. Because fracture and conduit permeability are a function of the cube and fourth power of aperture and hydraulic radius, respectively, they are very sensitive to larger fractures and dissolution features. Calibrated values of hydraulic conductivity from a numerical ground-water flow model are about 500 m d^{-1} and are much closer to values estimated from aquifer tests than from outcrop measurements. Although estimating permeability from outcrop measurements appears very limited, this technique has been used successfully for estimating stream-flow losses through fracture pavements in the Edwards aquifer (Zahm, 1998).

Permeability in karstic aquifers is often scale dependent (Teutsch & Sauter, 1991; Rovey, 1994; Huntoon, 1995). Therefore, measurements at smaller scales may not apply at larger scales. This is also true in the Edwards aquifer where it appears that permeability increases with scale (Halihan et al., this volume).

Differences between the different techniques indicate how the scale of measurement determines what is actually measured in a karstic aquifer. Small scale measurements such as permeability measured in core-plugs focus on the matrix of the aquifer and help quantify the volume of the water stored in the aquifer and relate dissolution to aquifer stratigraphy. Larger scale measurements such as transmissivity estimated from well tests or outcrop measurements focus on fractures and dissolution features that contribute to the larger-scale permeability and the transport porosity of the aquifer. However, even at this larger scale, major regional controls on permeability may be missed. In this case, larger-scale estimates of permeability from numerical models or springflow analysis should be used. Together, the different techniques help better characterize porosity and permeability in a complex karstic aquifer. These data document a dual porosity aquifer where most of the storage (porosity) is in the matrix and most of the flow (permeability) is in the fractures and conduits.

6 CONCLUSIONS

We used several different techniques to help define the porosity and permeability of a karstic aquifer. Each technique has its strengths and weaknesses and accounts for a different element of the porosity and permeability distribution in the aquifer. Core plugs are useful for defining the relationship between matrix porosity and permeability to stratigraphy and geophysical log response. Although estimates of porosity and permeability from geophysical log response are approximate, aquifer-wide trends in matrix porosity and permeability can be identified with the technique. Outcrop measurements are useful for estimating secondary porosity due to fractures and dissolution features. Although these measurements are likely biased toward higher values because of unloading effects and weathering, they are likely better than estimates based on core-plugs and poor-resolution geophysical logs. Permeability estimated from outcrop measurements of fractures and dissolution features are considerably higher than measured values because the technique assumes that all the measured features are linear and infinite in extent and generally should not be used to estimate permeability.

Well tests can be used to estimate well-scale permeability throughout the aquifer and generally measure the impact of fractures and perhaps conduits on aquifer permeability. However, care must be taken when using transmissivity estimated from well tests at a different scale as values at one scale may not apply at another. Also, because of the extreme heterogeneity in karstic aquifers, transmissivity estimated in a few wells should not be interpreted as representative of hydraulic properties in the area. Well tests can also be used to identify if transmissivity is spatially correlated as we found in the Edwards aquifer. This helps identify if different parts of the aquifer have generally similar hydraulic characteristics and can be used to guide the interpolation of results.

ACKNOWLEDGMENTS

This work was part of a study to characterize the regional porosity and permeability of the Edwards aquifer and was funded by the Edwards Underground Water District (now Edwards Aquifer Authority). We thank Erika Boghici and Norman Johns for assistance in data collection, entry, and analysis with the geographical information system; and Alan Dutton, Stephen Ruppel, Joseph Yeh, and Randall Marrett for assistance and comments during the studies. We also thank the Texas Water Development Board, the Edwards Underground Water District, and the Texas Natural Resources Information System for assistance and access to well files and Ken Bradbury and Steve Worthington for their thoughtful reviews. Publication authorized by the Director, Bureau of Economic Geology, The University of Texas at Austin.

REFERENCES

Alexander, K.B. 1990. Correlation of structural lineaments and fracture traces to water-well yields in the Edwards aquifer, Central Texas (M.A. Thesis). Austin: The University of Texas, 113 p.

Archie, G.E. 1942. The electrical resistivity log as an aid in determining some reservoir characteristics. *Transactions of the A.I.M.E.*, 146: 54-62.

Apostal, T.M. 1957. *Mathematical analysis – a modern approach to advanced calculus.* Reading, Massachusetts: Addison-Wesley, 553 p.

Barton, C.A. & Zoback, M.D. 1992. Self-similar distribution and properties of macroscopic fractures at depth in crystalline rock in the Cajon Pass scientific drill hole. *Journal of Geophysical Research*, 97: 5181-5200.

Brown, D.S., Petri, B.L. & Nalley, G.M. 1992. Compilation of hydrologic data for the Edwards aquifer, San Antonio area, Texas, 1991, with 1934-1991 summary. *Edwards Underground Water District, Bulletin* 51, 169 p.

Clark, Isobel, 1979. *Practical geostatistics:* London, Applied Science Publishers, Limited, 129 p.

Collins, E.W. & Hovorka, S.D. 1997. Structure map of the San Antonio segment of the Edwards aquifer and Balcones Fault Zone, South-Central Texas- structural framework of a major limestone aquifer- Kinney, Uvalde, Medina, Bexar, Comal, and Hays Counties: Bureau of Economic Geology Miscellaneous map no. 38, 14 p. + plate.

Cooper, H.H., Jr. & Jacob, C.E. 1946. A generalized graphical method for evaluating formation constants and summarizing well field history. *Transactions of the American Geophysical Union*, 27: 526-534.

Dahlquist, G. & Björck, Å. 1974. *Numerical methods.* Englewood Cliffs, New Jersey: Prentice-Hall, 573 p.

Deike, R.G. 1990. Dolomite dissolution rates and possible Holocene dedolomitization of water-bearing units in the Edwards aquifer, south-central Texas. *Journal of Hydrology*, 112(3-4): 335-373.

Ewing, T.E. 1991. Structural framework, in Salvador, A. (ed.) *The Gulf of Mexico basin:* Boulder, Colorado: Geological Society of America, J: 31-52.

Gillespie, J.R., Howard, C.B., Walsh, J.J. & Watterson, J. 1993. Measurements and characterisation of spatial distributions of fractures. *Tectonophysics*, 226: 114-141.

Groschen, G.E. & Buszka, P.M. 1997. Hydrogeologic framework and geochemistry of the Edwards aquifer saline-water zone, south-central Texas: U.S. Geological Survey Water-Resources Investigations Report 97-4133, 47 p.

Guyton, W.F. & Associates 1986. Drilling, construction, and testing of monitor wells for the Edwards aquifer bad-water line experiment: Austin, Texas, report to the City Water Board of San Antonio, the Edwards Underground Water District, Texas Water Development Board, and the U.S. Geological Survey, 56 p.

Halihan, T., Mace, R.E. & Sharp, J.M., Jr., this volume, Flow in the San Antonio segment of the Edwards aquifer- matrix, fractures, or conduits?

Harris, C., Franssen, R. & Loosveld, R. 1991. Fractal analysis of fractures in rocks – the Cantor's Dust method – a comment. *Tectonophysics*, 198: 107-115.

Hovorka, S.D., Dutton, A.R., Ruppel, S.C. & Yeh, Joseph 1996. Edwards aquifer ground-water resources-geologic controls on porosity development in platform carbonates, South Texas: Bureau of Economic Geology Report of Investigations no. 238, 75 p.

Hovorka, S.D., Mace, R.E. & Collins, E.W. 1995. Regional distribution of permeability in the Edwards Aquifer: Austin, Texas, Bureau of Economic Geology, The University of Texas, final report to the Edwards Underground Water District under contract no. 93-17-FO, 126 p.

Hovorka, S.D., Mace, R.E. & Collins, E.W. 1998. Permeability structure of the Edwards aquifer, South Texas – implications for aquifer management: Bureau of Economic Geology Report of Investigations no. 250, 55 p.

Huntoon, P.W. 1995. Is it appropriate to apply porous media ground water circulation models to karstic aquifers? in El-Kadi, A.I. (ed.) *Assessment of models for groundwater resource analysis and management*. Boca Raton, Louisiana: Lewis Publishers, p. 339-358.

Isaaks, E.H. & Srivastava, R.M. 1989. *An introduction to applied geostatistics*. New York, Oxford University Press, 561 p.

Klemt, W.B., Knowles, T.R., Elder G. & Sieh, T. 1979. Ground-water resources and model applications for the Edwards (Balcones Fault Zone) Aquifer in the San Antonio region, Texas: Texas Department of Water Resources Report 239, 88 p.

Mace, R.E. 1995. Geostatistical description of hydraulic properties in karst aquifers – a case study in the Edwards Aquifer: in Charbeneau, R.J. (ed.) *Groundwater Management – Proceedings of the International Symposium sponsored by the Water Resources Engineering Division*, American Society of Civil Engineers. p. 193-198.

Mace, R.E. 1997. Determination of transmissivity from specific capacity tests in a karst aquifer. *Ground Water*, 35: 738-742.

Maclay, R.W. 1995. Geology and hydrology of the Edwards aquifer in the San Antonio area, Texas: U.S. Geological Survey Water-Resources Investigations Report 95-4186, 64 p.

Maclay, R.W. & Small, T.A. 1976. Progress report on geology of the Edwards aquifer, San Antonio area, Texas, and preliminary interpretation of borehole geophysical and laboratory data on carbonate rocks: U.S. Geological Survey Open-File Report 76-627, 65 p.

Maclay, R.W. & Small, T.A. 1983. Hydrostratigraphic subdivisions and fault barriers of the Edwards aquifer, South-Central Texas, U.S.A. *Journal of Hydrology*, 61(1-3): 127-146.

Maclay, R.W. & Small, T.A. 1986. Carbonate geology and hydrology of the Edwards aquifer in the San Antonio area, Texas: Texas Water Development Board Report 296, 90 p.

Maclay, R.W., Small, T.A. & Rettman, P.L. 1980. Water-level, recharge, discharge, specific-capacity, well-yield, and aquifer test data for the Edwards aquifer in the San Antonio area, Texas: Texas Department of Water Resources Report LP-133, 83 p.

Maclay, R.W., Small, T.A. & Rettman, P.L. 1981. Application and analysis of borehole data for the Edwards aquifer in the San Antonio area, Texas: Texas Department of Water Resources Report LP-139, 88 p.

Marquardt, G.L. & Small, G.R. 1979. Records of wells, chemical analyses, and water levels of selected Edwards wells, Bexar County, Texas: Texas Department of Water Resources Report 237, 458 p.

Marrett, R. 1996. Aggregate properties of fracture populations. *Journal of Structural Geology*, 18: 169-178.

Myers, B.N. 1969. Compilation of results of aquifer tests in Texas: Texas Water Development Board Report 98, 532 p.

Murray, G.E. 1961. *Geology of the Atlantic and Gulf coastal province of North America:* New York: Harper Brothers, 692 p.

Poteet, Diane, Collier, Hughbert & Maclay, Robert 1992. Investigation of the fresh/saline-water interface in the Edwards aquifer in New Braunfels and San Marcos, Texas: Edwards Underground Water District Report 92-02, variously paginated.

Rasband, W. 1994. NIH Image, Version 1.50: U.S. National Institutes of Health.

Rose, P.R. 1972. Edwards Group, surface and subsurface, Central Texas: Bureau of Economic Geology Report of Investigations no. 74, 198 p.

Rovey, C.W. II 1994. Assessing flow systems in carbonate aquifers using scale effects in hydraulic conductivity: *Environmental Geology*, 24: 244-253.

Schultz, A.L. 1992. Using geophysical logs in the Edwards aquifer to estimate water quality along freshwater/saline-water interface (Uvalde to San Antonio, Texas): Edwards Underground Water District Report 92-03, 46 p.

Schultz, A.L. 1993. Defining the Edwards aquifer freshwater/saline-water interface with geophysical logs and measured data (San Antonio to Kyle, Texas): Edwards Underground Water District Report 93-06, 81 p.

Sharp, J.M., Jr. & Banner, J.L. 1997. *The Edwards aquifer, a resource in conflict:* GSA Today – A Publication of the Geological Society of America, 7(8): 1-9.

Sieh, T.H. 1975. Edwards (Balcones Fault Zone) aquifer test well drilling investigation: Texas Water Development Board unpublished file report, 127 p.

Smith, C.I. 1964. Physical stratigraphy and facies analysis, Lower Cretaceous limestones, Edwards Plateau, West Texas: Houston, Texas, Shell Development Company, EPR special report 45, 138 p.

Teutsch, G. & Sauter, M. 1991. Groundwater modeling in karst terranes – scale effects, data acquisition, and field validation: in *Proceedings of the Third Conference on Hydrology, Ecology, Monitoring, and Management of Ground Water in Karst Terranes, Nashville:* U.S. Environmental Protection Agency and National Ground Water Association, p. 17-34.

Theis, C.V. 1935. The relation between the lowering of the piezometric surface and the rate and duration of discharge of a well using groundwater storage: *Transactions of the American Geophysical Union*, 16: 519-524.

Turcotte, D.L. 1997. *Fractals and chaos in geology and geophysics*, 2nd edition: New York, Cambridge University Press, 398 p.

Woodruff, C.M., Jr. & Abbott, P.L. 1986. Stream piracy and evolution of the Edwards aquifer along the Balcones Escarpment, central Texas, in Abbott, P.L. & Woodruff, C.M., Jr. (eds) *The Balcones Escarpment – Geology, Hydrology, Ecology and Social Development in Central Texas:* Field trip guide book for the November, 1986, Geological Society of America Annual Meeting in San Antonio, Texas, p. 77-90.

Zahm, C.K. 1998. Use of outcrop fracture measurements to estimate regional groundwater flow – Barton Springs segment of Edwards aquifer, Central Texas (M.A. thesis): Austin, The University of Texas, 154 p.

Matrix, fracture and channel components of storage and flow in a Paleozoic limestone aquifer

STEPHEN R.H. WORTHINGTON & DEREK C. FORD
School of Geography and Geology, McMaster University, Hamilton, Ontario, Canada
GARETH J. DAVIES
Cambrian Ground Water, Tennessee, USA

ABSTRACT: Study of matrix, fracture, and channel characteristics in a limestone aquifer in the central Kentucky karst has highlighted the contrasts between the permeability components in this triple porosity aquifer. The rock matrix provides almost all the storage, but channels transmit almost all the flux of groundwater. However, the probability of a borehole intersecting a channel is only 0.004-0.03, so boreholes are not ideal for assessing flow in carbonate aquifers. The range in velocities between matrix and channel flow exceeds nine orders of magnitude. Two definitions of aquifer residence time can be made. The residence time of the water stored in the aquifer is dominated by matrix storage, and is some thousands of years. The residence time of recharge to the aquifer and of spring discharge from the aquifer is between four months and seven years, and is dominated by fracture and channel flow. Assessment of flow in carbonate aquifers is not simple because channels are the major permeability elements, yet the probability of a borehole intersecting one is estimated to be between only 0.4% and 3%.

1 INTRODUCTION

In carbonate and other fractured-rock aquifers there may be channels where groundwater flow can be rapid. The consequence is that aquifers with channeling deviate from simple porous medium or double porosity models. With flow in the rock matrix, fractures and channels it is appropriate to refer to such aquifers as triple porosity aquifers (Worthington, 1994; Quinlan et al., 1996). The purpose of this paper is to investigate the relative importance of these three porosity elements with respect to flow and storage in a carbonate aquifer.

Channels are defined as elongated voids having lengths at least ten times their diameter (Choquette & Pray, 1970). Research in Cornwall, Great Britain, and in the Stripa Mine, Sweden, has established that such channels may occupy 5-20% of a given fracture plane in granite (Tsang, 1993). However, in carbonate rocks some channels may be greatly enlarged by solution processes. Where apertures exceed 1 cm then turbulent flow commonly results, and such channels are often referred to as conduits (White, 1988, p. 292). Where channels are further enlarged to the size where people can enter them they are called caverns (Choquette & Pray, 1970, p. 224) or caves (White, 1988, p. 60; Ford & Williams, 1989, p. 242). Channels have an aperture range from < 1 mm to > 10 m.

In aquifers in carbonate rocks, the often rapid recharge of large volumes of meteoric water with low bicarbonate hardness results in dissolution and strong channeling that is focused in a small proportion of the fracture planes. This is a consequence of two factors. The first is the non-linear nature of carbonate dissolution; as thermodynamic equilibrium is approached the rate decreases by several orders of magnitude (Plummer & Wigley, 1976), with the result that groundwater is slightly under-saturated with respect to calcium (or magnesium) carbonate at most sites where there is notable flow. The second factor is the positive feedback relationship between dissolution rate and discharge which permits larger channels to grow at the expense of smaller ones (Ford & Williams, 1989, p. 249 et seq.). The two factors combine to create broadly dendritic networks of channels (Palmer, 1991). In unconfined carbonate aquifers in moist climates, channeling should always develop.

It is not realistic to expect to encounter representative channel groundwater by means of sampling randomly drilled wells due to the low probability of intersecting any channel from the surface, even when drilling at intuitively logical locations. Consequently, other sampling methods are necessary, such as from adits or springs. In recent years the effects of channeling have been a concern at certain proposed high-level nuclear waste disposal sites, and these effects have been investigated by injecting tracers (usually fluorescent dyes) into boreholes and recovering them in underlying adits excavated for the purpose (Abelin et al., 1991a, b; Birgersson et al., 1993). In carbonate rocks numerous measurements have been made in channel networks, either directly within caves that are of enterable dimensions or at natural springs discharging from them (White, 1988; Ford & Williams, 1989; White & White, 1989).

The central Kentucky karst around Mammoth Cave is developed in limestones and includes a mature triple porosity aquifer that is exploited by a large number of water wells. There are also many explored caves, including Mammoth Cave, which now has more than 500 km of mapped, inter-connected passages and is the most extensive known cave in the world (Courbon 1989). There is a considerable body of knowledge on the large channels that make up the cave and on their hydrogeology, both from studies within the cave and from measurements at springs (Palmer, 1981; White & White, 1989). The presence of both wells and mapped caves make this aquifer ideal for investigating processes of flow through limestones because many porosity elements can be studied.

The aquifer is in the St. Louis, Ste. Genevieve and Girkin Formations (Mississippian), which consist of sparites and micrites with minor oolitic limestones, shaley limestones, silty calcarenites and dolomitic horizons, and has an aggregate thickness of about 120 m. The strata dip gently to the north-west, (Palmer, 1981; Hess et al., 1989), where the deeply entrenched Green River receives most of the discharge from the aquifer. Turnhole Spring is one example of the large springs found on the south bank of the river. It discharges the flow from an area of ~217 km^2 (Fig. 1); its catchment has been the subject of a number of studies and is one of the best defined in central Kentucky (Quinlan & Ewers, 1989).

The central Kentucky karst has a total area of about 1500 km^2. In Turnhole and adjacent karst catchments, Quinlan and Ray (1981) measured water levels in some 1500 wells and carried out more than 500 tracer tests in order to determine flow paths and water-table elevations. The water table in Figure 1 is based primarily on their water level meas-

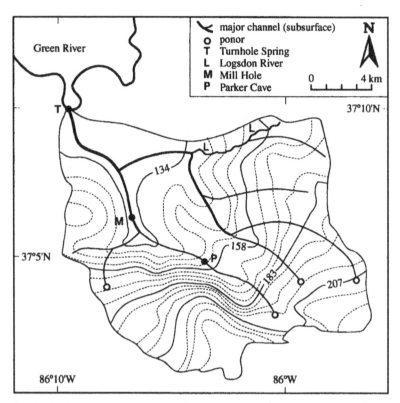

Figure 1. Map of the Turnhole Spring catchment, showing the convergence of flow to major subsurface channels. The contours are water table elevations, with a contour interval of 20 ft (6.1 m) (after Quinlan & Ray, 1981).

urements in 200 wells in the catchment. The water table generally slopes to the north-west, but locally there are troughs which coincide with flow along large channels. Some of these channels can be entered within caves such as Parker Cave and Mammoth Cave. The longest accessible channel segment, known as 'Logsdon River', is in Mammoth Cave (Fig. 1). Mill Hole and Cedar Sink are 'karst windows' created by collapses from the surface into underlying cave channels that are unknown but must be large; in the windows there is surface flow for short distances (Quinlan et al., 1986; Quinlan & Ewers, 1989).

Hess & White (1989) determined that almost all the discharge from this area into the Green River is by channel flow. The numerous groundwater tracing studies, spring studies and cave studies in the central Kentucky karst attest to the importance of flow in large channels. However, there are also more than 4000 water wells in the karst, almost all of which must derive their flow from flow in fractures, in the matrix, and in small channels because they are most unlikely to have intercepted large channels. The three components of matrix, fracture and channel flow will be discussed separately below to offer insights into the hydraulic functioning of the aquifer.

2 MATRIX POROSITY AND PERMEABILITY

Brown & Lambert (1963) tested core samples from the Ste. Genevieve Formation, measuring a total porosity of 3.3%, an effective porosity of 2.4%, and a hydraulic conductivity of 2×10^{-11} m/s. We have measured an average matrix porosity of 1.4% from samples of micritic limestone in the Ste. Genevieve Formation at a quarry close to Mill Hole, the site of special investigations outlined below. These low porosity values are within the range of average values for Paleozoic and other well-lithified carbonates worldwide (Bonacci, 1987, p. 47); in this aquifer, porosity of the matrix is low, not nearly as high as has been sometimes inferred in major hydrogeology texts, and does not support the concept that karstic limestones necessarily have high porosity (Freeze & Cherry, 1979, p. 37).

3 FRACTURE POROSITY AND PERMEABILITY

There are six boreholes of 150 mm or 200 mm diameter spaced at intervals of ~100 m across the inferred line of the major channel immediately upstream of Mill Hole (Quinlan et al., 1986). Slug tests were performed on these and a further three boreholes elsewhere in the catchment. Water depths ranged from 9 m to 24 m. The data were analyzed using the method of Hvorslev (1951). Hydraulic conductivity results ranged from 4×10^{-6} to 1×10^{-4} m/s, with an arithmetic mean 3×10^{-5} m/s and a geometric mean of 6×10^{-6} m/s. The permeability due to fracturing is likely to fall between these two latter values. The geometric mean would represent the average permeability of the fractures if they were discontinuous or open for only short distances, and the arithmetic mean would represent the average permeability of the fractures if they were continuous. In this region bedding planes may be open for considerable distances, so the fracture permeability is likely to be closer to the arithmetic mean than the geometric mean.

Down-hole videos in the six boreholes at Mill Hole showed that the spacing of fractures with visible enlargement was 0.7 m, but major open fractures with visible apertures of 2-12 cm had an average spacing of 8.5 m. These fractures all appeared to be bedding planes. The porosity due to fracturing can be calculated by using the cubic law

$$K = \left(\rho g N b^3 / (12\mu) \right) \tag{1}$$

where K is the hydraulic conductivity, ρ is fluid density, g is the acceleration due to gravity, N is the distance between fractures, b is fracture aperture, and μ is dynamic viscosity (Snow, 1968).

Figure 2 shows the relationship between hydraulic conductivity, porosity, fracture aperture and fracture spacing for a set of smooth parallel fractures, with pure water at 10°C, based on Equation 1. Using the ranges of fracture hydraulic conductivity and fracture spacing determined above, then the porosity due to bedding planes is 0.006% to 0.05% and fracture apertures are in the range 0.2-0.7 mm (Fig. 2). These calculated hydraulic apertures are for smooth fractures, and actual rough-walled fractures would be somewhat larger. However, most of the observed major open fractures, with openings of 2-12 cm, probably represent local enlargements such as vugs or drilling enlargements.

In the low-dip, inter-basin setting of central Kentucky, bedding planes provide the principal fractures for horizontal groundwater movement. Continuous and prominent parting planes are visible in rock outcrops, and their importance is demonstrated by the

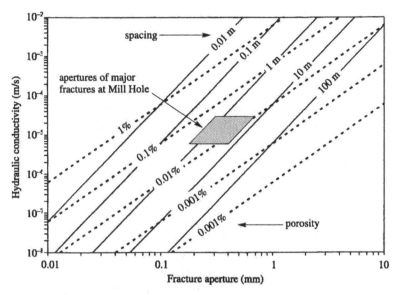

Figure 2. Relationship between hydraulic conductivity and fracture aperture, spacing, and porosity for a set of parallel fractures with pure water at 10°C, showing the range of apertures of bedding planes at Mill Hole.

fact that the great majority of the passages in Mammoth Cave are developed along them (Palmer, 1981). Joints and faults are likely to only add marginally to the overall or gross fracture porosity of the limestone, though they are important for vertical infiltration in the unsaturated zone.

4 CHANNEL POROSITY AND PERMEABILITY

Two methods have been used to calculate channel porosity. Ashton (1966) proposed that pre-storm water of moderate hardness (i.e. channel water) would be flushed out of springs by lower hardness storm water recharge by a piston flow process. Atkinson (1977) used this technique to estimate reasonable volumes of the major channels in the central Mendip Hills, England, an area of complex geological structure (Waltham et al., 1997). In the central Kentucky karst, continuous measurement of electrical conductivity at Cedar Sink for a period of more than one year showed that there were sharp drops in conductivity following major storms (Hess & White, 1988). The average lag time between the onset of precipitation and the arrival of low conductivity water at the Sink was found to be 35 hours. It can be assumed that the low conductivity water represents the concentrated recharge from allogenic streams sinking at the limestone contact, most of which sink ~ 11 km upstream of Cedar Sink. The average length of the entire groundwater catchment between the divide and Turnhole Spring is 17 km, suggesting an average travel time of 54 hours from the upstream limits. The water discharged from Turnhole Spring during a period of 54 hours at high stage thus represents an estimate of the aquifer channel volume.

There are no discharge records for Turnhole Spring because the water discharges directly into the Green River and thus is difficult to gauge independently. However, the total discharge for the central Kentucky karst can be calculated from the difference between the discharge at upstream and downstream gauging stations (Hess & White, 1989). The 989 km^2 aggregate catchment between the two stations is predominantly karst terrain, draining to a number of major springs along the river. It is reasonable to assume that the discharge from Turnhole Spring during the first 54 hours is broadly in proportion to its 217 km^2 catchment area, and thus is 217/989 of the total flow from the 989 km^2 catchment. The estimated average storm discharge for the first 54 hours of major storm events in the period monitored by Hess & White (1988) was 62 m^3/s for the whole 989 km^2 catchment, which gives a mean channel volume of 2.6 × 10^6 m^3 for the Turnhole basin.

An alternative method to calculate channel volume is to use estimates of average runoff and average channel velocity. Hess & White (1989) used discharge records from the Green River to calculate an average specific discharge of 19.5 L/s/km^2 from the central Kentucky karst area. Figure 3 shows channel velocities from 2877 tracer tests between sinking streams and springs. The data come from tests in carbonate rocks in 37 countries which are published in the literature; they show a close to log-normal velocity distribution and have a median velocity of 0.022 m/s. Repeated tracer tests along a 6 km channel from Parker Cave to Mill Hole gave velocities ranging from 0.0025 m/s at low flow to 0.13 m/s at high flow (Quinlan & Ewers, 1989). The geometric mean of these two values is 0.018 m/s, which is close to the global average of 0.022 m/s. Thus flow velocities in major channels in the Turnhole Spring catchment are similar to the global velocity distribution.

The wetted cross-section of the major channels can now be calculated by

$$A = CR / v \qquad\qquad (2)$$

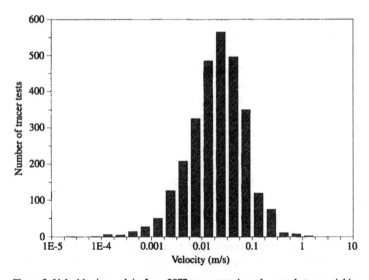

Figure 3. Velocities in conduits from 2877 tracer tests in carbonates between sinking streams and springs.

where A is the channel cross-section, C is the catchment area of the channel, R is the run-off, and v is average channel velocity. A total of 34 flow paths were traced in the Turn-hole Spring catchment by Quinlan & Ray (1981), of which the major ones are shown in Figure 1. Using Equation (2) the calculated volume of the channels along the 34 flow paths is $2.0 \times 10^6 \, \text{m}^3$.

It should be noted that Equation (2) uses the straight-line distance between tracer injection and tracer recovery points. Worthington (1991) calculated an average sinuosity of 1.48 from a sample of channels in 96 accurately mapped caves. This suggests that the true average velocity along such channels is ~0.032 m/s (i.e. 0.022 m/s × 1.48) and that the value of the channel cross-section in Equation (2) is correspondingly less. The decreased channel cross-section is compensated by the increased channel length, however, so that the calculated channel volume is unchanged.

Both estimates of channel volume are subject to considerable error. Williams (1983) has argued that the higher-hardness water displaced by a flood represents not only channel water but also water from the epikarst (subcutaneous zone). Furthermore, Bakalowicz (1979) and Meiman et al. (1988) have shown that the lag between rise in water level in channels and the drop in electrical conductivity varies with antecedent moisture conditions, as would be expected. The second calculation only represents the channels along 34 flow paths that have been traced. These probably represent most of the largest channels in the catchment, but there must also be numerous smaller ones which will increase the aggregate channel volume. However, despite the possible errors, the similarity of our two independent calculations lends credibility to an actual channel volume of approximately $2.3 \times 10^6 \, \text{m}^3$.

In the Mammoth Cave region, the effective thickness of the saturated zone can be estimated from cave maps. There has been very little scuba diving in flooded segments of the channels but in one cave, Hidden River, the stream flows up a vertical fissure from a bedding plane channel that is 17 m below the water surface (Quinlan & Ewers, 1989, p. 77). In abandoned passages the deepest channel flow that has been identified is to at least 23 m below a former water table in Mammoth Cave (Palmer, 1989, p. 313). For a saturated zone thickness of 23 m and a karst area without surface drainage of 159 km² in the Turnhole Spring catchment, the channel porosity is $2.3 \times 10^6 \, \text{m}^3 / 3.66 \times 10^9 \, \text{m}^3$, or 0.06% of the volume of the aquifer.

The average flow in channels can be calculated by considering a cross-section through the aquifer midway between the divide and the springs, as shown in Figure 4. Based on a specific discharge of 19.5 L/s/km² (Hess & White, 1989), the mean discharge through this section will be 3.16 m³/s. From the water table map of Quinlan & Ray (1981), the average hydraulic gradient in this region of the aquifer is 0.004. Assuming an effective saturated zone thickness of 23 m and a matrix hydraulic conductivity of 2×10^{-11} m/s, then Darcy's Law can be used to calculate a matrix flow discharge through the cross-section of 2×10^{-8} m³/s. By similar calculations fracture flow is 0.007-0.03 m³/s, depending on whether the arithmetic or geometric mean hydraulic conductivity of the slug test data is used. Channel flow represents the remaining 3.12-3.153 m³/s, i.e. between 99.0% and 99.8% of the flow through this representative cross-section of the aquifer.

An 'equivalent hydraulic conductivity' for channel flow may be calculated by rearranging Darcy's Law, which gives a value of 3×10^{-3} m/s. This is an average value across the mid-section of the aquifer, and is only an approximation. Hydraulic gradients along the major channels are mostly lower than average hydraulic gradients, but Logsdon

River, one of the four main channels at the cross-section, has open-channel flow and a steeper gradient than the two largest channels at the cross-section (Fig. 4). Furthermore, flow in channels is likely to be turbulent at times, so a non-linear flow law such as the Darcy-Weisbach equation would be needed for detailed analysis of channel flow. However, the use of an equivalent hydraulic conductivity for channel flow does facilitate comparison with other flow components and with other aquifers.

A simple method of calculating the proportions of flow through the aquifer is to use the ratios of the hydraulic conductivity values. For matrix flow this may be calculated by

$$F_m = K_m / \left(K_m + K_f + K_c \right) \tag{3}$$

where F is the fraction of flow, K is hydraulic conductivity, and the subscripts m, f, and c refer to the matrix, fractures, and channels, respectively. The proportion of flow that takes place through fractures and channels may be calculated in a similar way.

Mean hydraulic gradients along major channels were determined for the three largest catchments in the water table map of Quinlan & Ray (1981), which has a contour interval of 20 ft (6.1 m). For each channel shown on the map, the hydraulic gradient (i) and catchment area (A) were determined for each channel segment between contours. Regression of the resultant data gave

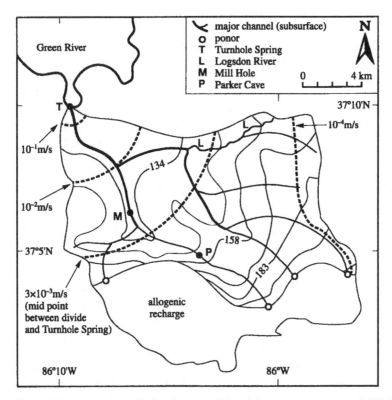

Figure 4. Equivalent channel hydraulic conductivities increasing in a downstream direction in the Turnhole Spring catchment.

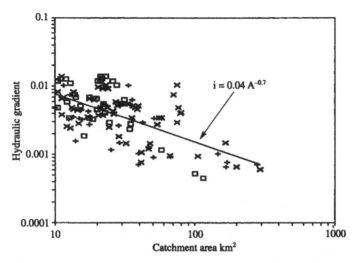

Figure 5. Decreasing hydraulic gradients in a downstream direction along channels in the Turnhole Spring catchment (•) and adjacent Bear Wallow (×) and Graham Springs (+) catchments.

$$i = 0.039A^{-0.70} \tag{4}$$

Results are shown in Figure 5. Hydraulic gradients diminish in a downstream direction. The exponent of –0.70 is within the range –0.37 to –0.83 determined by Flint (1974) for a number of rivers. Such exponents have been explained in terms of minimizing energy expenditure in the channel network (Richards, 1982, p.227). Conduits within carbonate bedrock are similar in many respects to surface stream channels, such as being organized into dendritic tributary networks and usually carrying a significant clastic sediment load. Thus it is not surprising that trends in gradients are similar. Nevertheless, there are some fundamental differences for channels within the bedrock: closed-channel flow is common and since the conduits are within the bedrock channel width cannot easily adjust in response to changing hydraulic conditions.

The trend of decreasing hydraulic gradients in a downgradient direction shown in Figure 5 is the opposite to water table gradients in an ideal porous medium. In that case increasing gradients are needed to drive the increasing flow in a downgradient direction. Thus regional trends in hydraulic gradients provide a diagnostic tool for identifying aquifers where channeling is important.

In the Turnhole Spring catchment decreasing gradients downstream are accompanied by an increase in discharge and a reduction in catchment width. These three factors combine to increase the equivalent hydraulic conductivity (Fig. 4).

5 DISCUSSION AND CONCLUSIONS

The porosity and permeability data of the Turnhole aquifer are summarized in Table 1. The most striking feature is that almost all of the storage is in the rock matrix, but channels account for almost all of the permeability. The magnitude of the contrasts in porosity and permeability between the three permeability elements mean that the results are

Table 1. Matrix, fracture and channel porosity and permeability in the turnhole spring catchment.

Flow mode	Porosity (%)	Fraction of storage (%)	Hydraulic conductivity (m/s)	Fraction of permeability (%)
Matrix	2.4	95.6-97.3	2×10^{-11}	0.0000007
Fracture	0.006-0.05	0.2-2.0	6×10^{-6}-3×10^{-5}	0.2-1
Channel	0.06	2.4	3×10^{-3}*	99-99.8

Note: * 'Equivalent' hydraulic conductivity since channel flow is likely to be turbulent

relatively insensitive to the errors in the measurements. These errors could be as great as +/– 50% in some cases but do not affect the main conclusion. These findings are quite similar to those of Atkinson (1977), who studied the flow components in the Mississippian limestone in the Mendip Hills, England, though without the benefit of well data; the channels there were estimated to hold only 3% of groundwater storage but to account for an some 60-80% of the flow.

An understanding of groundwater movement in the aquifer is facilitated by the subdivision of permeability into three different categories, each of which can be measured independently. However, the three-fold division used above must still be considered a simplification. For instance, the difference between total matrix porosity and effective matrix porosity emphasizes that there may be a broad range of matrix residence times. Similarly, with fractures and channels there is a continuum from narrow, poorly-connected tectonic fractures with laminar flow to well-connected conduits with turbulent flow, which results in another broad range of velocities and residence times. Furthermore, in addition to storage within the bedrock there is also storage with the soil zone above the bedrock (Williams, 1983).

The porosity and permeability contrasts in the Turnhole Spring catchment and other limestone aquifers have interesting implications for flow velocities, for aquifer residence times, and for aquifer testing using boreholes.

5.1 Flow velocities

The contrasts in permeability suggest that solute transport in the aquifer is best considered by analyzing the different flow components rather than seeking average values by lumping all components together. This is well illustrated by the contrasts between tracer test results where injections have been into conduits and into boreholes. Velocities along conduits typically range from hundreds of meters to kilometers per day (Fig. 3). The tracers are usually injected into dolines or sinking streams, both of which represent the upstream end of conduits. The dendritic nature of the channeling determines that tracers injected in any conduit within the Turnhole Spring catchment will rapidly move to the spring. This has been a concern at Mammoth Cave National Park because spills on highways such as Interstate Highway 65 may drain into dolines, and then move quickly through conduits into caves within the Park. As a result there have been detailed studies to establish where any such spills would drain (Quinlan & Ray, 1981; Meiman et al., 1996).

Velocities of flow in fractures are considerably slower than in the conduits. For example, Recker et al. (1988) determined an average velocity of 47 m/day from two tracer tests to Mill Hole from a well 300 m away. Fracture velocity may be approximated by

$$v = Ki / n_e \tag{5}$$

which is based on Darcy's Law, where i is the hydraulic gradient and n_e is the effective porosity. Assuming a hydraulic gradient of 0.004 and the values in Table 1, then Equation (5) gives fracture velocities of 7-38 m/day.

Matrix flow velocities are much slower than this. Given the hydraulic gradient of 0.004 and the values in Table 1, Equation (5) gives matrix flow velocities of 0.1 mm/year. A calculated 96-97% of the total amount of groundwater in the Turnhole aquifer is stored in the rock matrix but with such low velocities it is essentially immobile when compared to the flow in fractures and channels.

5.2 *Aquifer residence times*

An average residence time for groundwater in the carbonate aquifer can be calculated by considering the three components of matrix, fracture and channel flow. However, in this triple porosity medium there are two definitions of residence time

$$T_r = T_m R_m + T_f R_f + T_c R_c \tag{6}$$

$$T_s = T_m S_m + T_f S_f + T_c S_c \tag{7}$$

where T is residence time, R is recharge to the aquifer, S is the storage in the aquifer, the subscripts m, f, and c refer to matrix, fracture and channel, respectively, T_r is the residence time of water recharging the aquifer, and T_s is the residence time of the water within the aquifer. The three components of R and S are dimensionless fractions, the sum of which are both unity.

Residence time of aquifer recharge. Approximately 28% of the catchment has surface runoff which enters the main carbonate aquifer via sinking streams (Fig. 4). An additional 16% of the catchment has a sandstone caprock. Within Mammoth Cave there are a large number of vertical shafts close to the edge of the caprock (Hess et al., 1989), demonstrating that much of the caprock runoff also enters directly into conduits. The remaining 56% of the catchment has a limestone surface with many dolines. Rapid recharge to the aquifer occurs via macropore flow through the soil into channels in the limestone and also by direct recharge at the base of some dolines where soil is absent. The fraction of this area of limestone with rapid recharge is uncertain, but may represent up to 50%, especially after heavy precipitation.

A number of studies have been made of aquifer recharge via dolines and these show that there are a number of flow paths with a range of velocities (Gunn, 1983; Zambo & Ford, 1997). Friederich & Smart (1981) and Kogovšek (1997) were able to measure tracer velocities, making injections at the base of dolines and monitoring tracer arrival at drip points in caves 40-100 m below the surface. Tracer arrival times at the monitoring points ranged from 8 minutes to several days. The rapid downward flow from these doline experiments is the result of flow down channels.

These three modes of direct recharge to the channel network in the carbonate aquifer account for an estimated 45-70% of the recharge to the aquifer. The residence time of this channel flow is the product of mean residence time from conduit traces (153 s/m from the data set in Fig. 3) and the mean conduit path of 11 kilometers, which gives 19 days.

The remaining 30-55% of recharge passes through the soil to enter fractures and the matrix. The mean annual recharge in the limestone area is 615 mm (Hess & White, 1989) and, assuming an average soil moisture content of 300 mm, the mean soil water residence time then is about six months. The vertical hydraulic conductivity in fractures is not precisely known, but is likely to be one to two orders of magnitude less than the horizontal hydraulic conductivity, which suggests that a value around 10^{-7} m/s is appropriate. Given a matrix hydraulic conductivity of 2×10^{-11} m/s, the ratio between matrix and fracture hydraulic conductivity is ~1:5000. Consequently, one part in 5000 of the 30-55% of the matrix plus fracture recharge will enter the matrix, or about 0.01% of total recharge. The remaining 29.99-54.99% of recharge enters fractures.

For fracture flow the residence time in both the unsaturated zone and the saturated zone must be considered. Bottrell & Atkinson (1992) used tracers to study velocities in fractures in the unsaturated zone from the surface down to drip points in cave passages located 50-100 m below the surface. Monitoring of tracer arrival in the cave showed that fracture residence times varied from days to years. The range of velocities suggest it is not possible at present to determine an accurate mean residence time in fractures in the unsaturated zone for recharge water, but average times of weeks to several years are possible.

Average fracture velocities in the saturated zone can be calculated using Equation (5), which gives a range of 7-38 m/day. Fracture flow in the saturated zone will feed into channels after average fracture flow path lengths that may be tens to hundreds of meters. Assuming an average path of 100 m then the fracture-flow residence time in the saturated zone is probably between two days and two weeks. If soil residence times are added to fracture flow times in both the saturated and unsaturated zones then the mean residence time of fracture flow is likely to be between seven months and ten years.

The hydraulic conductivity of the matrix is several orders of magnitude less than the hydraulic conductivity of the fractures, so matrix velocities are extremely slow. However, the pathways between fractures are fairly short (1-10 m). Matrix velocities of 0.1 mm/year were calculated earlier, so residence times may average 10,000 to 100,000 years.

A summary of the residence times of water recharging the aquifer are given in Table 2. Using Equation (6) the mean residence time is in the range 1-15 years.

Residence time of aquifer storage. The mean residence time of water stored in the aquifer is calculated by Equation (7), and uses the previously calculated data for storage (Table 1) and for residence times of the three flow components (Table 2). Results are give in Table 2. Calculated mean residence times are between 9600 years and 96,000 years.

The two methods of calculating residence times give results that differ by more than three orders of magnitude. Water sampled at Turnhole Spring will have the same residence time as recharge water, whereas sampling from boreholes is likely to under-sample the channel component and have greater ages. Extraction of matrix water from rock samples will yield the highest ages. Furthermore, the diffusion characteristics of any natural

Table 2. Residence times for the turnhole spring carbonate aquifer.

Flow mode	Fraction of recharge (%)	Residence time (years)	Fraction of storage (%)
Matrix	0.01	10,000-100,000	95.6-97.3
Fracture	30-55	0.6-10	0.2-2
Channel	45-70	0.05	2.4

Note: From Equation (6) mean residence time for recharge (T_r) is between one and 15 years. However, from Equation (7) mean residence time for storage (T_s) is between 9,600 and 96,000 years.

or artificial tracer will have an effect on groundwater ages measured with it. Thus the measurement of residence time in carbonates is not straightforward.

5.3 *Probability of boreholes intercepting channels*

It is well known that channels account for most of the effective permeability in the Turn-hole limestone aquifer, and we have shown how its importance can be estimated quantitatively. A most important question is to determine whether a small number of randomly-located boreholes are likely to facilitate representative aquifer sampling. This is because the standard method of assessing groundwater and solute movement is to drill several boreholes and assume that they will intercept the major permeability elements of an aquifer. This assumption is so fundamental that it is not even mentioned in most hydrogeology texts. Consequently, an absence of evidence for the presence of channeling is taken as evidence of an absence of channeling. However, several lines of evidence show this premise is false.

Dissolutional channel networks normally are composed of a hierarchical dendritic pattern, with increasing channel size in the downgradient direction. This is demonstrated by the tributary pattern of flow shown on Figure 1, and by observations in Mammoth Cave and in thousands of other caves in carbonate rocks. One method of assessing the spatial coverage of channels within the Turnhole Spring catchment is to consider the 34 conduit flow paths that were subject to tracer testing by Quinlan & Ray (1981). Measurements of accessible (relict) passages in Mammoth Cave show that such channels typically have width: height ratios of about 3:1 (White & Deike, 1989). If their form is approximated by a rectangle and a sinuosity of 1.5 is assumed (as explained above), then the calculated spatial coverage of these 34 Turnhole conduits is 5.7×10^5. The conduits are located within the 159 km^2 of the catchment that is without surface drainage, and thus underlie only ~0.36% of its area.

A second method is to utilize the maps of Mammoth Cave, which are based on detailed measurements. Crystal Cave is a section of Mammoth Cave with a very dense network of cave passages. A total of 26,790 m of interconnecting galleries have been mapped, with a spatial coverage of 9.4×10^4 m^2 (A.N. Palmer, 1998, personal communication). The cave underlies a rectangular surface area of 1200 m by 2350 m, and thus occupies 3.3% of it.

The two calculations of 0.36% and 3.3% bracket the likely spatial coverage of channels. The first calculation is an underestimate because it only includes the 34 major flow paths that have been detected in the Turnhole catchment, and there must be many more, smaller, channels. The second calculation is for an unusually dense network of caves, where a succession of passages were used in turn by the cave streams. These were left abandoned in the unsaturated zone as the water table dropped and new, lower conduits were formed (Palmer, 1981).

A third method of assessing channeling is to use downhole video data. The six boreholes at Mill Hole were drilled close to one of the largest traced channels in the Turnhole catchment (Fig. 1). Video logs in 334 m of open hole in the six wells revealed a number of fracture enlargements but none exceeded 12 cm in height. Furthermore, the slug test data showed values which are quite standard for Paleozoic carbonate aquifers, with the highest hydraulic conductivity being 1×10^{-4} m/s. In sum, the video logs and slug tests gave no indication of the presence of a major channel nearby.

These three sets of measurements all point to there being a very low probability of a borehole intercepting a channel. Thus if no channels were found in the 4-40 boreholes typically drilled for site assessments, this would not provide evidence that channeling was absent at the site. This is demonstrated for instance by the failure of a drilling program in the Maligne Lake valley, Alberta. Tracer tests have demonstrated that one of the largest karst channels in North America must underlie this valley (Smart, 1988). Following an electrical resistivity survey, a total of 2800 m of drilling was undertaken in the limestone to attempt to intercept this channel, but the largest cavity detected was only 5 cm in height (Ford & Fish, 1971).

The low probability of boreholes intercepting channels is one of the major problems in assessing flow in carbonate aquifers. From borehole tests at Mill Hole there is no evidence of the presence of channeling, yet the boreholes are close to a channel that has a diameter of several meters. Worthington & Ford (1995) have outlined several borehole testing and monitoring strategies that may suggest the presence of channels. Tracer testing between boreholes is the most promising of these methods, but there remains no proven strategy for detecting the presence of channeling using a small number of boreholes. Consequently, the level of uncertainty in interpreting the hydrogeology of carbonate aquifers is much higher than in porous media such as sands.

ACKNOWLEDGMENTS

J.F. Quinlan was an enthusiastic contributor to the early stages of this research, and his untimely death is deeply regretted. We thank J. Meiman and M. Ryan of Mammoth Cave National Park for assistance in the field work and for open discussion of ideas, C.C. Smart for the loan of equipment, A.N. Palmer for sharing cave statistics, and W.B. White and P.W. Williams for useful review comments. This research was partially supported by a National Science and Engineering Council of Canada research grant to D.C. Ford.

REFERENCES

Abelin, H., L. Birgersson, J. Gidlund & I. Neretnieks 1991a. A large scale flow and tracer experiment in granite. 1. Experimental design and flow distribution. *Water Resources Research*, 27: 3107-3117.

Abelin, H., L. Birgersson, L. Moreno, H. Widen, T. Agren & I. Neretnieks 1991b. A large scale flow and tracer experiment in granite. 2. Results and interpretation. *Water Resources Research*, 27: 3119-3135.

Ashton, K. 1966. The analyses of flow data from karst drainage systems. *Transactions of the Cave Research Group of Great Britain.*, 7: 161-203.

Atkinson, T.C. 1977. Diffuse flow and conduit flow in limestone terrain in the Mendip Hills, Somerset (Great Britain). *Journal of Hydrology*, 35: 93-110.

Bakalowicz, M. 1979. Contribution de la géochimie des eaux à la connaissance de l'aquifère karstique et de la karstification. Thèse de doctorat d'état ès sciences naturelles, Université Pierre et Marie Curie, Paris, 269 p.

Birgersson, L., Moreno, L., Neretnieks, I., Widen, H. & Agren, T. 1993. A tracer migration experiment in a small fracture zone in granite. *Water Resources Research*, 29: 3867-3878.

Bonacci, O. 1987. Karst hydrology. Berlin: Springer-Verlag, 184 p.

Bottrell, S.H. & Atkinson, T.C. 1992. Tracer study of flow and storage in the unsaturated zone of a karstic limestone aquifer, in:, Hötzl, H. & Werner A. (eds) *Tracer hydrology*. Rotterdam: Balkema, p. 207-211.

Brown, R.F. & T.W. Lambert 1963. Reconnaissance of ground water resources in the Mississippian Plateau region, Kentucky: United States Geological Survey Water Supply Paper 1603, 58 p.

Choquette, P.W. & Pray, L.C. 1970. Geological nomenclature and classification of porosity in sedimentary carbonates: *American Association of Petroleum Geologists Bulletin*, 54: 207-250.

Courbon, P., Chabert, C., Bosted, P. & Lindsley, K. 1989. Atlas of the great caves of the world. St Louis, Cave Books, 369 p.

Flint, J.J. 1974. Stream gradient as a function of order, magnitude and discharge. *Water Resources Research*, 10: 969-973.

Ford, D.C. & Fish, J.E. 1971. Geo-electrical resistivity surveys in the lake Maligne valley, Jasper National Park. Contract report for Parks Canada, 27 p.

Ford, D.C. & Williams, P. 1989. Karst geomorphology and hydrology. London: Unwin Hyman, 601 p.

Freeze, R.A. & J.A. Cherry 1979. *Groundwater*. Englewood Cliffs, NJ: Prentice-Hall, 604 p.

Friederich, H. & Smart, P.L. 1981. Dye trace studies of the unsaturated-zone recharge of the Carboniferous Limestone aquifer of the Mendip Hills, England, in Beck, B.F. (ed.) *Proceedings, International Congress of Speleology, 8th, Bowling Green:* Huntsville, National. Speleological Society, p. 283-286.

Gunn, J. 1983. Point-recharge of limestone aquifers – a model from New Zealand karst. *Journal of Hydrology*, 61: 19-29.

Hess, J.W. & White, W.B. 1988. Storm response of the karstic carbonate aquifer of southcentral Kentucky. *Journal of Hydrology*, 99: 235-252.

Hess, J.W. & White, W.B. 1989. Water budget and physical hydrology, in White, W.B. & White, E.L. (eds) *Karst hydrology: Concepts from the Mammoth Cave area*. New York: Van Nostrand Reinhold, p. 105-126.

Hess, J.W., Wells, S.G., Quinlan, J.F. & White, W.B. 1989. Hydrogeology of the south-central Kentucky karst, in White, W.B. & White, E.L. (eds) *Karst hydrology: Concepts from the Mammoth Cave area*. New York: Van Nostrand Reinhold, p. 15-63.

Hvorslev, M.J. 1951. Time lag and soil permeability in ground water observations: U.S. Army Corps of Engineers Waterways Experimentation Station, Vicksburg, Mississippi Bulletin 36, 30 p.

Kogovšek, J. 1997. Water tracing tests in vadose zone, in Kranj, A. (ed.) *Tracer hydrology 97*. Rotterdam: Balkema, p. 167-172.

Meiman, J., Ewers, R.O. & Quinlan, J.F. 1988. Investigation of flood pulse movement through a maturely karstified aquifer at Mammoth cave national Park: a new approach, in *Proceedings of the second conference on environmental problems in karst terranes and their solutions, Nashville, Tennessee, 1988*, National Water Well Association, Dublin, Ohio, p. 227-262.

Meiman, J.M, Leech Jr., B.T., Fry, J.F. & Ray, J.A. 1996. *Groundwater hazard map of the Turnhole Spring karst groundwater basin:* Mammoth Cave, Mammoth Cave National Park, 50 p.

Palmer, A.N. 1981. *A geological guide to Mammoth Cave National Park:* Teaneck, New Jersey: Zephyrus Press, 196 p.

Palmer, A.N. 1989. Stratigraphic and structural control of cave development and groundwater flow in the Mammoth Cave region, in White, W.B. & White, E.L. (eds) *Karst hydrology: Concepts from the Mammoth Cave area*. New York: Van Nostrand Reinhold, p. 293-316.

Palmer, A.N. 1991. Origin and morphology of limestone caves. *Geological Society of America Bulletin*, 103: 1-21.

Plummer, L.N. & Wigley, T.M.L. 1976. The dissolution of calcite in CO_2-saturated solutions at 25° C and 1 atmosphere total pressure. *Geochimica et Cosmochimica Acta*, 40: 191-202.

Quinlan, J.F. & R.O. Ewers 1989. Subsurface drainage in the Mammoth Cave area, in White, W.B. & White, E.L. (eds) *Karst hydrology: Concepts from the Mammoth Cave area.* New York: Van Nostrand Reinhold, p. 65-103.

Quinlan, J.F. & Ray, J.A. 1981. Groundwater basins in the Mammoth Cave Region, Kentucky: Occasional Publication #1, Friends of the karst, Mammoth Cave.

Quinlan, J.F., Ewers, R.O. & Palmer, A.N. 1986. Hydrogeology of Turnhole Spring groundwater basin, Kentucky, in *Geological Society of America Centennial Field Guide – Southeastern Section,* p. 7-12.

Quinlan, J.F., Davies, G.J., Jones, S.W. & Huntoon, P.W. 1996. The applicability of numerical models to adequately characterize ground-water flow in karstic and other triple-porosity aquifers, in Ritchey, J.D. & Rumbaugh, J.O. (eds) *Subsurface fluid-flow (groundwater) modelling: West Conshohocken, American Society for Testing and Materials,* Special Technical Paper 1288, p. 114-133.

Recker, S.A., Ewers, R.O. & Quinlan, J.F. 1988. Seepage velocities in a conduit-adjacent porosity system of a karst formation and their influence on the movement of contaminants, in *Proceedings of the second conference on environmental problems in karst terranes and their solutions, Nashville, Tennessee, 1988.* National Water Well Association, Dublin: Ohio, p. 265-287.

Richards, K. 1982. *Rivers:* London, Methuen, 358 p.

Snow, D.T. 1968. Rock fracture spacings, openings and porosities. Journal of the Soil Mechanics and Foundations Division, *Proceedings, American Society of Civil Engineers,* 94: 73-91.

Smart, C.C. 1988. Quantitative tracing of the Maligne karst system, Alberta, Canada. *Journal of Hydrology,* 98: 185-204.

Tsang, C.F. 1993. Tracer transport in fracture systems, in Bear, J., Tsang, C-F. & de Marsily, G. (eds) *Flow and contaminated transport in fractured rock.* San Diego: Academic Press, p. 237-266.

Waltham, A.C., Simms, M.J., Farrant, A.R. & Goldie, H.S. 1997. *Karst and caves of Great Britain.* London: Chapman and Hall, 358 p.

White, W.B. 1988. *Geomorphology and hydrology of karst terrains.* New York: Oxford University Press, 464 p.

White, W.B. & Deike, G.H. 1989. Hydraulic geometry of cave passages, in White, W.B, & White, E.L. (eds) *Karst hydrology: Concepts from the Mammoth Cave area.* New York: Van Nostrand Reinhold, p. 223-258.

White, W.B. & White, E.L. 1989. *Karst hydrology: Concepts from the Mammoth Cave area.* New York: Van Nostrand Reinhold, 346 p.

Williams, P.W. 1983. The role of the subcutaneous zone in karst hydrology. *Journal of Hydrology,* 61: 45-67.

Worthington, S.R.H. 1991. Karst hydrogeology of the Canadian Rocky Mountains (PhD thesis). McMaster University, 380 p.

Worthington, S.R.H. 1994. Flow velocities in unconfined carbonate aquifers. *Cave and Karst Science,* 21: 21-22.

Worthington, S.R.H. & Ford, D.C. 1995. Borehole tests for megascale channeling in carbonate aquifers, in *Proceedings, 26th Congress of the International Association of Hydrogeologists, Edmonton, Alberta, June 5th-9th 1995,* unpaginated.

Zambo, L. & Ford, D.C. 1997. Limestone dissolution processes in Beke doline, Aggtelek National Park, Hungary: *Earth Surface Processes and Landforms,* 22: 531-543.

Flow in the San Antonio segment of the Edwards aquifer: matrix, fractures, or conduits?

TODD HALIHAN & JOHN M. SHARP JR.
Department of Geological Sciences, The University of Texas at Austin, Austin, Texas, USA

ROBERT E. MACE
Bureau of Economic Geology, The University of Texas at Austin, University Station, Austin, Texas USA

ABSTRACT: Understanding water movement through fractured karst aquifers is difficult, but it is important for those dependent upon these resources. Much of central Texas depends primarily on the fractured and karstified San Antonio segment of the Edwards aquifer. The problem of scale dependent permeability makes interpretations of the aquifer difficult by limiting our ability to effectively utilize small-scale data for larger scale problems. We used permeabilities measured at the small- (lab and outcrop), well-, and regional-scales to evaluate if small-scale data can be used to reproduce and interpret well- and regional-scale data. Small-scale data sets of matrix permeability, fracture aperture, and conduit size from cores and outcrops were utilized. Well-scale permeabilities were estimated from pump tests. Regional-scale permeabilities were estimated from numerical models. A modified layered aquifer model was used to calculate well- and regional-scale permeabilities from the small-scale data. Using an average regional hydraulic gradient to compute Reynolds numbers, the small-scale permeability data were used to predict the occurrence of non-linear laminar to turbulent flow. The results indicate that, in general, fractures control flow on the well-scale in the Edwards, and that many wells sample non-linear laminar to turbulent flow within the aquifer. The results also indicate that conduits are not the major contributors to well-scale permeability in the Edwards, but control regional-scale permeabilities and have turbulent flow. Finally, the results indicate that pump tests would yield no measurable drawdown in approximately 15% of wells due to either fractures or conduits.

1 INTRODUCTION

The ease with which water flows through an aquifer, quantified as permeability, can be difficult to determine. Permeability ranges over at least 17 orders of magnitude (Freeze & Cherry, 1979, p. 29; Neuzil, 1994). This variation in permeability is a key factor to understanding groundwater flow and solute transport in the subsurface (Rehfeldt et al., 1989).

In fractured karst aquifers, the difficulty in quantifying permeability is amplified by flow in fractured and dissolution zones. This is illustrated by the fact that measured permeability generally increases with scale in fractured karst aquifers (Kiraly, 1975). This is

129

generally referred to as the scale effect. Permeability measurements in a karstic aquifer can vary by as much as nine orders of magnitude (Maclay & Land, 1988; Hovorka et al., 1995) as shown in Figure 1.

In the San Antonio segment of the Edwards aquifer (hereafter referred to as the Edwards aquifer), quantifying permeability is important because of the increasing demands on the aquifer which require users to understand how the aquifer functions so that it can be properly utilized and protected (Sharp & Banner, 1997). Fortunately, the Edwards aquifer has a large permeability database available over a range of scales from core samples to regional model estimates. Utilizing these data, this research addresses the question of what controls flow in the Edwards aquifers: the matrix, the fractures, or the conduits. The question is not trivial, as workers in the Edwards aquifer have widely varying opinions of what is the appropriate conceptual model of the aquifer.

2 PREVIOUS WORK AND DEFINITIONS

Previous work has investigated the scale effect of permeability from field data and theoretical interpretations. In order to define scales in hydrology, different nomenclatures have evolved. The nomenclature used in this paper is similar to that of Bradbury & Muldoon (1990). Small scale refers to permeameter tests, fracture measurements, or conduit measurements that take place in the laboratory or outcrop and generally make a measurement over a volume of 0.01 m^3 to 10 m^3. This would correlate with the laboratory scale of Dagan (1986). Well scale refers to well or packer tests that occur on a scale of $100\text{-}1000 \text{ m}^3$, which is similar to the local scale of Dagan (1986). Finally, the regional scale refers to volumes of greater than 1000 m^3. Dagan (1986) uses the same term.

Kiraly (1975) first noticed the scale effect in karst aquifers of Switzerland, in which permeability continually increased from the small- to the regional-scale. He hypothesized that the increase from small- to well- scale was caused by fractures, and that the largest permeabilities on the regional-scale were caused by karstic conduits. Quinlan et al. (1992) compiled over 1800 dye traces from 25 countries, and concluded that average flow velocity values continually increased with scale. Extending the work of Brace (1984), Clauser (1992) noted that permeability increased approximately three orders of magnitude from the small-scale to the well-scale in crystalline rocks. However, Clauser suggested that the permeability of fractured media measured on the well-scale would not continue to increase when measured on the regional-scale. Other authors have also suggested that if a sufficiently large volume of rock was selected, a single, representative value for permeability could be determined (Long et al., 1982; Odling, 1997). Rovey (1994) examined carbonate aquifer permeability with variogram models. He found that for fractured unkarstified carbonates, a range could be determined in which the permeability reached a constant value. But, for mature, well-developed karst aquifers, he suggested that permeability instead increased to 'practical infinity'.

Models of the scale effect depend on which theory is used. Does the value of permeability reach a constant value at an appropriate scale, or does it increase indefinitely? Previous research has not attempted to combine the various scales of permeability with a quantitative, physical model. In fractured media, permeability data can be obtained on the matrix and fractures. In karstic media, data can also be collected on conduit sizes. Few, if any, models of which we are aware have tested the scale effect by incorporating these various data to determine which heterogeneities control the permeability of a fractured karst aquifer. A quantitative physical understanding of the scale effect makes quantifying permeability more robust, enabling the proper physical model to be used. An aquifer could be assessed on many scales, and inconsistencies could be investigated by collecting appropriate data.

Definitions for matrix, fractures, and conduits vary widely. For this paper, terminology was based upon observable hydrogeologic features in outcrops and caves. The terminology, which may provide a useful standard, is similar to that of Atkinson (1985), (see also Ford & Williams, 1989, pp. 166-170), however, the terminology is based purely on the aquifer's geometric characteristics and not upon the potential or interpreted flow regime. In this paper, matrix refers to rock where no fractures visible to the unaided eye are present. Fracture refers to fractures that are visible to the unaided eye that are open, not filled by minerals. Conduit refers to dissolution features that extend for a range greater than that of the well-scale and are visible to the unaided eye. Dissolution feature refers to solutional features which are limited to meter-scale or smaller. For modeling, fractures are two-dimensional uniform slots and conduits are one-dimensional pipes.

3 SITE DESCRIPTION

Hydrostratigraphic relationships for the Edwards aquifer are given in Rose (1972), Maclay & Small (1986), Pavlicek et al. (1987), and Sharp & Banner (1997). The aquifer consists of Cretaceous limestones and dolomites that have undergone multiple periods of karstification. It is not well understood whether the high permeabilities observed in the aquifer are due to heterogeneities in the matrix, well-connected fractures, karstified conduits, or some combinations thereof. The aquifer supplies one of the highest naturally flowing wells with a discharge of 1.58 m^3 s^{-1} (25,000 gpm) (Swanson, 1991). It also has springs that discharge from fractures in the outcrop at combined average rates between 5 and 15 m^3 s^{-1} (Sharp & Banner, 1997).

4 PERMEABILITY DATABASE

Permeability data are available for the Edwards aquifer on several scales (Fig. 1). Small-scale data are available from core samples and measurements of fracture apertures and conduits in outcrop. Pump test data evaluate the well-scale. Finally, three numerical models constructed for the Edwards aquifer provide estimates of regional-scale permeability. Permeability data collected on any scale have errors associated with those data, and we refer the reader to the data sources for discussions of errors.

In this study, the distributions of matrix permeability, fracture apertures, and conduit sizes were approximated as lognormal distributions. The distributions of fracture aperture and conduit sizes can also be modeled as power law distributions. By choosing lognormal distributions, the largest apertures and the largest conduits may not be represented in the analysis, resulting in a conservative estimate of the highest permeabilities.

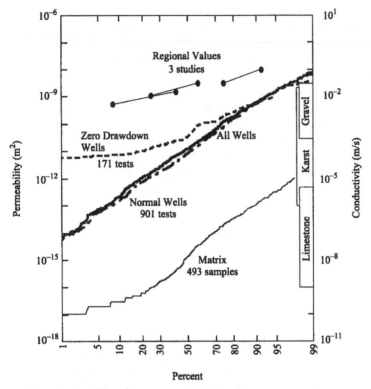

Figure 1. Permeability scale effect for the San Antonio segment of the Edwards aquifer. Cumulative distribution function for permeability values measured from cores (Hovorka et al., 1993; Hovorka et al., 1995; Hovorka, personal communication), wells (Hovorka et al., 1995; Mace, 1997), and regional model (Klemt et al., 1979; Maclay & Land, 1988; Thorkildsen & McElhaney, 1992) data. Zero drawdown wells are wells in which no measurable drawdown occurred, but 0.3 m of drawdown (resolution of the test data collected by drillers) was assumed in order to calculate the permeability. Permeability values for 3 regional modeling studies were calculated using a range of thickness (110-230 m) to convert from transmissivity to permeability. Permeability ranges for limestone, karst, and gravel (Freeze & Cherry, 1979) are illustrated for comparison. The individual data points for matrix and well data are not indicated because the quantity of data available provides nearly continuous lines.

4.1 *Small-scale*

Small-scale data for a fractured karst aquifer such as the Edwards are difficult to obtain. Representative samples of the matrix can be difficult to collect from cores. Fracture data can usually only be collected from the surface at outcrops or creek beds, or from caves and tunnels in the subsurface. Conduit data are yet more difficult to obtain due to the difficulty in determining the length of conduits that are smaller than can be explored. For the Edwards aquifer, many small-scale data are available, but additional small-scale data are yet needed for quantifying flow.

Matrix. The matrix permeability data utilized for this study consist of 493 measurements from 0.025 m (1 inch) core plugs taken from various studies (Fig. 1). 195 measurements are from Hovorka et al. (1993); an additional 195 measurements are from Hovorka et al. (1995); and the remaining 103 matrix permeability measurements are from unpublished measurements (Hovorka, personal communication). The matrix permeability appears to have a bottom truncation at a lower value of approximately 10^{-17} m^2. For this study, the matrix permeability was modeled as a lognormal distribution with a geometric mean of 1.3×10^{-15} m^2 (Fig. 2). A value one standard deviation above the mean yielded a matrix permeability of 4.6×10^{-14} m^2, and a value of 3.7×10^{-17} m^2 resulted for one standard deviation below the mean.

Fractures. Fracture permeability was calculated from fracture apertures measured on three transects of roadcuts of the Edwards aquifer (Hovorka et al., 1998). A total of 776 fracture apertures were measured with a feeler gauge and a metal ruler, primarily in sub-vertical fractures (Fig. 2). Intrinsic permeability was calculated from apertures using the cubic law (Lamb, 1932), which estimates the permeability of a fracture, $k_f[L^2]$, as:

$$k_f = \frac{b^2}{12} \tag{1}$$

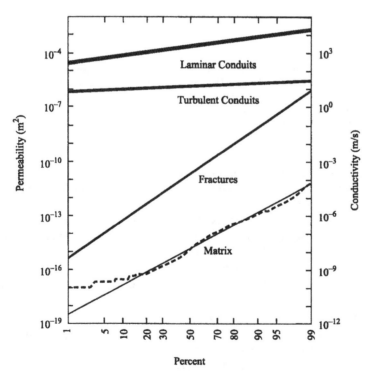

Figure 2. Small scale permeability information for the San Antonio segment of the Edwards aquifer. Data for 493 matrix measurements (dashed line) illustrated along with lognormal model for matrix measurements (thin solid line). Lognormal distributions of hydraulic conductivity for fractures and conduits were obtained using aperture and diameter data shown in Figure 3. Fracture permeability (medium solid line) was calculated using Equation (1). Laminar and turbulent conduit hydraulic conductivity (heavy solid lines) were calculated using Equations (2) and (4), respectively.

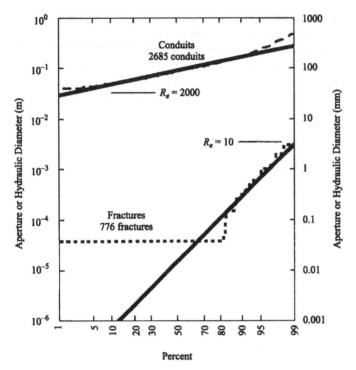

Figure 3. Conduit and fracture size distribution. Fracture data shown for 776 fracture aperture measurements illustrated along with lognormal model of distribution. Uniform distribution of fractures below the 80th percentile is due to the measurement limit. Hydraulic diameter data shown for 2685 conduits. Reynolds numbers are shown for non-linear laminar ($10 < R_e < 2000$) transition for fractures and turbulent ($R_e > 2000$) transition for conduits calculated using estimates of hydraulic gradient (0.11-0.42 m km^{-1}) for the Edwards aquifer (McKinney & Watkins, 1993).

where b is the fracture aperture [L]. The limit of the feeler gauge technique is apparent in the data set (Fig. 3), with the data truncated below the 80th percentile. Fracture apertures were assumed to follow a lognormal distribution (Figs 2 and 3). The geometric mean fracture aperture used was 0.0155 mm. This results in a geometric mean permeability of 2×10^{-11} m^2. A value one standard deviation above the mean yielded a fracture permeability of 2×10^{-9} m^2, and a value of 2×10^{-13} m^2 resulted for one standard deviation below the mean. It is not assumed that all fracture apertures measured at the outcrop were well connected, but it is assumed that the apertures of well-connected fractures follow the measured outcrop distribution.

The fracture distribution was tested for the validity of laminar Darcian flow. Reynolds numbers were calculated using the 0.11-0.42 m km^{-1} range of hydraulic gradients observed in the Edwards (McKinney & Watkins, 1993). A Reynolds number of 10 was used to represent the onset of non-linear laminar flow in the fractures (Lindquist, 1933; Scheidegger, 1974, pp. 152-187; Fetter, 1994, pp. 143-144). This occurred with apertures greater than 3.1-4.8 mm (Fig. 3). Full turbulence was estimated at a Reynolds number of 2000 (Streeter, 1948). This occurred in fractures that ranged in aperture from 18-28 mm. From these calculations, it was estimated that the flow was laminar for 99% of the fractures.

Conduits. Conduit permeability was calculated as the cross-sectional diameter of conduits measured at seven outcrops (Hovorka et al., 1995). They estimated the diameter of 2685 conduits (Figs 2 and 3) by tracing conduits from photomosaics onto mylar sheets, and then scanning and interpreting these sheets using National Institutes of Health image processing software (Rasband, 1994). It cannot be expected that all of the conduits will maintain the same diameters along their entire length, but we assume that the extended conduits follow the measured distribution.

The permeability of the conduits was calculated from the Darcy-Weisbach equation assuming both laminar and smoothly turbulent flow. The laminar hydraulic conductivity of the conduits, K_l [L t^{-1}], was calculated using:

$$K_l = \frac{g}{\nu}\frac{d^2}{32} \tag{2}$$

where d is the hydraulic diameter [L], g is the gravitational constant [L t^{-2}], and ν is the kinematic viscosity [L^2 t^{-1}] (Gupta, 1989, pp. 549-551). The turbulent hydraulic conductivity (derived from Turcotte & Schubert, 1982, pp. 239-240; Halihan et al., 1998), K_t [L t^{-1}], was calculated for smoothly turbulent flow using the empirical relationship between friction factor and Reynolds number:

$$f = 0.3164 R_e^{-1/4} \tag{3}$$

where f [–, dimensionless] is the friction factor and R_e [–] is the Reynolds number. Substituting the definition for friction factor and Reynolds number yields an expression for the turbulent hydraulic conductivity of:

$$K_t = 4.706\frac{g^{4/7}}{\nu^{1/7}}(\frac{d}{2})^{5/7}. \tag{4}$$

Intrinsic permeability was calculated by dividing the hydraulic conductivities by $g\,\nu^{-1}$. (Note: For non-Darcian flow, not only does the intrinsic permeability formulation change, but the gradient is no longer linearly proportional to the flux. Equation (4) only accounts for conduit geometry and fluid properties.)

Reynolds numbers calculated using hydraulic gradients for the Edwards aquifer indicate that flow would be turbulent ($R_e > 2000$) in conduits larger than 0.025-0.039 m. This includes 95-99% of the conduits (Fig. 3). The conduit-size distribution was modeled with a lognormal distribution, with a geometric mean conduit having a hydraulic diameter of 0.087 meters. This yields a hydraulic diameter of 0.139 m for a value one standard deviation above the mean, and a value of 0.054 m for one standard deviation below. This yields a median laminar permeability of 2.4×10^{-4} m^2, and a median turbulent permeability of 1.4×10^{-6} m^2 (Fig. 2).

The question arises using this conduit distribution as to whether the largest conduits have been included. The conduits may follow a power law or fractal distribution that would yield larger conduit diameters, and cave maps in the Edwards indicate meter scale conduits are common (Mace, in press; Veni, 1988). In order to test the effect of large conduits on permeability, we included both 1-meter and 10-meter conduits for analysis.

4.2 *Well-scale*

Well-scale data are abundant for the Edwards aquifer (Fig. 1), but a major difficulty with some of the data is created by high permeabilities. A number of well tests in the aquifer recorded no measurable drawdown (Fig. 4). These wells have been classified as zero drawdown wells (Hovorka et al., 1995; Mace, in press). The other well tests that had measurable drawdown have been termed 'normal' well tests for this study.

Normal well tests. Data on well permeability for 901 normal well tests in the Edwards aquifer (Fig. 1) were utilized (Hovorka et al., 1995). The normal well tests were obtained throughout the aquifer, but the majority were in the confined portion. Some tests were standard pumping tests, but most were specific capacity tests. Permeability was estimated from the specific capacity data using an empirical relationship derived for the Edwards aquifer (Mace, 1997). The geometric mean permeability of the normal well tests was 8.64×10^{-12} m^2 with a value of 1.97×10^{-10} m^2 representing one standard deviation above the mean, and a value of 3.79×10^{-13} m^2 representing one standard deviation below.

Well test data also included information on aquifer thickness and the open interval (most well completions are open hole) for the well tests. These parameters were modeled as normal distributions for this study. The formation had an average thickness of 170 m with a standard deviation of 30 m. The open interval had an averaged length of 100 m with a standard deviation of 30 m. Although the deviation for the open interval approximates much of the data, it was used as an average comparison against the full formation thickness model, maintaining the same standard deviation.

Zero drawdown well tests. Approximately 15% of the well test data collected by Hovorka et al. (1995) had no measurable drawdown reported (Figs 1 and 4). If valid, these 171 tests would yield an infinite permeability. Some researchers believe these tests are unusable or useless. Others claim that permeability may be at some 'practical infinity' (e.g., Rovey, 1994). Specific capacity for these well tests was estimated by assuming a drawdown of 0.3 m (1 foot), which was considered the resolution of the test data collected by drillers (Hovorka et al., 1995; Mace, in press). The transmissivity was then estimated using an empirical relationship derived for the Edwards (Mace, 1997). The geometric mean permeability of the zero drawdown well tests was 5.42×10^{-11} m^2 with a value of 3.01×10^{-10} m^2 representing one standard deviation above the mean, and a value of 9.75×10^{-12} m^2 representing one standard deviation below the mean.

The uncertainty in these tests may eliminate them as worthwhile data, but ignoring them adds bias. These data probably represent the highest permeabilities in the aquifer. Therefore, they must be considered. The average pumping rate for these well tests was 0.04 m^3 s^{-1} (625 gpm) with the highest pump rate at 0.525 m^3 s^{-1} (8337 gpm). The average borehole diameter for these tests was 0.23 m (9 inches) with the largest being 1.52 m (60 inches) (Hovorka et al., 1995).

4.3 *Regional-scale*

Regional estimates of permeability for the Edwards aquifer are limited to a few modeling studies (Fig. 1) (Klemt et al., 1979; Maclay & Land, 1988; Thorkildsen & McElhaney, 1992). Klemt et al. (1979) used a number of values for transmissivity for their model of

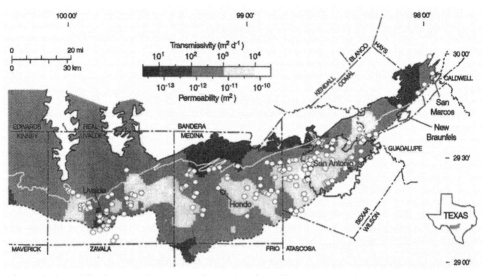

Figure 4. Map of the San Antonio segment of the Edwards aquifer. Location of zero drawdown wells (wells with no measurable drawdown during pumping) is shown with white circles. Smoothed kriging of the transmissivity field included for comparison with locations of zero drawdown wells (Mace, in press; Hovorka et al., 1988). Legend for transmissivity includes estimate of permeability using average aquifer thickness of 170 meters to convert values. Solid dark line indicates northern physical boundary for the San Antonio segment of the Edwards. Dashed line indicates the bad water line which forms the southern chemical boundary for the aquifer. The white line indicates the recharge boundary which approximates the confined/unconfined boundary for the aquifer.

the aquifer, but the highest regional value used was 2.9 $m^2 s^{-1}$. Using a range of thickness equal to two standard deviations in formation thickness (i.e., 110 m to 230 m), the highest regional permeability was calculated as 1×10^{-9} to $3 \times 10^{-9} m^2$. Maclay & Land (1988) use a slightly higher value for their highest transmissivity of 9.3 $m^2 s^{-1}$. Given the possible range of aquifer thickness, this equates to a regional permeability of 3×10^{-9} to $9 \times 10^{-9} m^2$. Finally, Thorkildsen & McElhaney (1992) use one of the smallest values for the high transmissivity area of their model. Their highest value was 1.4 $m^2 s^{-1}$, equating to regional permeabilities of 5×10^{-10} to $1.5 \times 10^{-9} m^2$.

5 PERMEABILITY COMBINATION MODELS

To evaluate the effects of different high permeability heterogeneities on the scale effect in the Edwards aquifer, models were tested that combine the matrix permeability distribution with the fracture and conduit permeability distributions (Fig. 5). These models are steady state and are modifications of equations for layered aquifers. In addition to the assumptions used for layered aquifers, four additional assumptions are necessary:

1. the matrix permeability, fracture apertures, and conduit diameters follow a lognormal distribution;

2. the cubic law is valid for fractures and Equations (2) and (3) are valid for laminar and turbulent flow in conduits;

3. the regions of high permeability do not strongly affect the matrix, and

4. the distributions of well-connected fracture apertures and conduit sizes are equivalent to the measured distributions of fracture apertures and conduit diameters.

5.1 *Matrix + horizontal fractures*

The effects of multiple fractures were examined to test the effect of fracture density on effective permeability. Effective permeability of matrix and fractures can be calculated using:

$$k_e = k_m - k_m (\frac{\Sigma b_f}{b_t}) + (\frac{\Sigma b_f^3}{12 b_t}) \qquad (5)$$

where k_e is the effective permeability (L^2), k_m is the matrix permeability (L^2), b_f is the fracture aperture (L), and b_t is the total thickness of the aquifer (L). Equation (4) can be solved in terms of the fracture permeability by substituting $(12 k_f)^{1/2}$ for b_f, where k_f is the fracture permeability (L^2). This is a modified form of permeability of a layered aquifer (Leonards, 1962; Fetter, 1994).

The k_e distribution was estimated from the distribution of k_m, b_f, and b_t using Monte Carlo simulations with 10,000 trials per simulation (Jensen et al., 1997, pp. 61-64). These simulations were performed using the distribution for the full aquifer thickness, and the open interval length for b_t. Each fracture was generated separately for the simulations. These simulations were performed for models of 1, 10, 50, and 100 horizontal fractures.

5.2 *Matrix + horizontal conduit*

The model for combining the matrix with the conduit permeability assumed a conduit would typically intersect a well horizontally. The effective permeability for the matrix and a conduit was calculated using:

$$k_e = k_m + (k_c - k_m)(\frac{d_c}{b_t}) \qquad (6)$$

where d_c is the hydraulic diameter of the conduit $[L]$, and k_c is the laminar or turbulent permeability of the conduit $[L^2]$. This calculation is an average over the thickness of the aquifer (b_t), and is restricted laterally to a width equal to d_c. The k_e distribution was estimated from the distribution of k_m, k_c, d_c, and b_t using Monte Carlo simulations. These simulations were performed with four different models; laminar models with b_t as the total formation thickness and one with b_t as the open interval of the wells, and corresponding turbulent models.

The above conduit model is sufficient for investigating small- and well-scale effects since the flow regime near the conduit would be linear. However, on a regional basis, two questions remain unanswered: First, how would large diameter conduits (1-10 meters) observed in the Edwards affect the estimated regional model permeabilities (see Veni, 1988, for conduit examples)? Secondly, how does the averaging of low permeability areas adjacent to the conduit affect the calculated value? In order to answer these questions values for a 1 meter and a 10 meter diameter conduit were calculated using Equation (5).

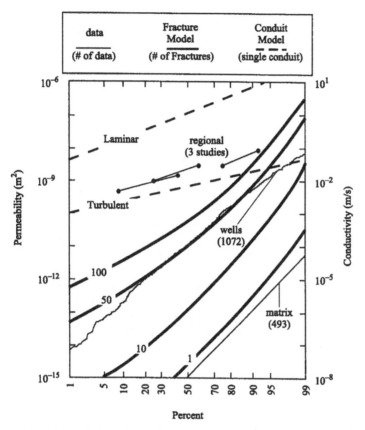

Figure 5. Monte Carlo models for matrix + multiple horizontal fractures and matrix + single horizontal conduit combination models. Matrix, well, and regional distributions shown for comparison.

In addition, permeabilities were calculated for a 10 km wide, 170 m thick region using the average well permeability for k_m to estimate the effect upon regional scale permeabilities.

6 RESULTS

The results indicate that large changes in the effective permeability are created by high permeability heterogeneities. The models generated for the aquifer result in permeabilities that have a range greater than the observed permeability data.

6.1 *Matrix + horizontal fractures*

For the multiple horizontal fracture model using the full aquifer thickness, large increases in permeability are evident over the matrix values. While a single fracture increased effective permeability by only 2.3 times at the median, 100 fractures can increase the median effective permeability by over four and a half orders of magnitude (Fig. 5). This would equate to an average connected fracture density of 1 fracture every 1.7 meters, with the majority of the fractures smaller than can be measured with a standard feeler gauge. Using the open interval of the wells instead of the full aquifer thickness had little effect on the value of k_e.

The best fit of the multiple horizontal fracture model to the well test data occurred with 50 fractures (average connected fracture density of 1 fracture every 3.4 meters). At this density, the model fit approximately 80% of the well test data with an error in the median values of 8% (Fig. 5). The model predicted higher values for the effective permeability below the 10th percentile and above the 90th percentile.

To determine the relative effects of the 50 horizontal fractures and the matrix on effective permeability, 20 trials of the model were examined. In the random trials, the matrix contributed less than 0.5% of the permeability in all but one case. The largest number of fractures contributing greater than 1% of the total effective permeability was 6 of 50 fractures, which occurred in 4 of the 20 trials analyzed. These four trials had an effective permeability range of 1.9×10^{-13} to 6.0×10^{-12} m^2. The largest fracture had an aperture of 1.77 mm contributing 46.2% of the permeability. The smallest number of fractures contributing greater than 1% of the permeability was 1 of 50, which occurred in 6 of the 20 trials. These six trials had an effective permeability range of 1.3×10^{-12} to 1.3×10^{-8} m^2 with the largest single fracture of 29.0 mm contributing 99.9% of the permeability. These limited tests indicate that flow occurs in a limited number of fractures for the model, and that some of these fractures are flowing under non-linear laminar ($10 < R_e < 2000$) to turbulent ($R_e > 2000$) flow conditions.

6.2 *Matrix + horizontal conduit*

The laminar horizontal conduit model resulted in a predicted effective permeability increase of over 7 orders of magnitude compared to the matrix alone (Fig. 5). This prediction is significantly above the calculated well test permeabilities and the regional permeabilities. The turbulent horizontal conduit model resulted in a predicted effective permeability increase of over 5 orders of magnitude. It is not expected that every well in the Edwards intersects a conduit of significant dimensions, so it is appropriate to compare the horizontal conduit model prediction either with the upper end of the well permeability distribution or with the regional values. The median value of the turbulent model is 20 times larger than the median well test value. However, these values are similar to some of the calculated values for the zero drawdown wells. The matrix and conduit model accommodating partial penetration using the open interval of the well for the total thickness is only 1.8 times higher than the model for the formation thickness.

The large conduit (1 and 10 meter) models yielded laminar permeabilities of 2×10^{-4} and 2×10^{-1} m^2 for the laminar model and 5×10^{-8} and 2×10^{-6} m^2 for the turbulent model, respectively. When averaged across a 10 km section of aquifer, the model yields laminar permeabilities of 1×10^{-8} and 1×10^{-4} m^2 and turbulent permeabilities of 1×10^{-11} and 2×10^{-9} m^2, respectively. Thus, while a 1 meter conduit occupies only 0.00005% of the flow area in this estimate, it would contribute 99.94% of the flow under laminar conditions over a 10 km wide segment of aquifer.

7 DISCUSSION

When examining carbonate rock outcrops, fractures and karst conduits can commonly be observed. The permeability of these features can be estimated as easily as the matrix permeability (Fig. 3). When examining permeabilities on the well- or regional-scale however, the location of permeable features is not simple to determine. For the Edwards aquifer, facies analysis indicates the presence of high permeability matrix in some locations (Hovorka et al., 1995). Lineament studies that delineate fracture traces in terms of contours instead of discrete fracture traces indicate the Edwards aquifer is dominated by fracturing (Wermund et al., 1978). Up to 27 meter high (90 ft) bit drops observed in drilling wells, caliper tests that indicate large dissolution features, and live blind catfish discharged from wells all strongly indicate the presence of conduits (Longley, 1981; Thornhill et al., 1988; Poteet et al., 1992). So which heterogeneities control flow in the Edwards aquifer, and what additional data might be needed to improve our understanding of the aquifer?

7.1 *Does the matrix control the Edwards' permeability?*

In the Edwards aquifer, the matrix does not appear to contribute significantly to permeability at either the well-scale or regional-scale (Fig. 6, Model A). From the permeability data, the average matrix can account for only two well permeabilities and none of the regional values. Even if the entire aquifer was composed of the highest 1% of the matrix permeability (an unrealistic assumption), less than 38% of the well data could be explained. Additionally, if the matrix could explain the well- or regional-scale data, the contributions of the fractures and conduits would have to be negligible, which is not likely.

7.2 *Do the fractures control the Edwards' permeability?*

For the Edwards aquifer, the fractures can explain the majority of the observed well permeability data (Figs 5 and 6, Model B or C). With a relatively simple model, fractures can be combined with the matrix to yield a match to the well permeability distribution. For individual wells, only a small number of fractures in the model contribute significantly to flow. This observation is made in many fractured systems (Dyke, 1995; Marrett, 1996). While fracture apertures are generally small, a few larger well-connected millimeter scale apertures can contribute a great deal to permeability in an aquifer.

The departures of the well test data at low permeabilities from the 50 horizontal fracture model can be explained by a number of factors (Fig. 5). For example, these wells may not intersect many fractures. For economic reasons, some wells were only drilled until sufficient discharge was obtained. This could result in a decreased well test permeability because the majority of the permeability in the wells appears to be due to a small number of fractures.

The highest permeability well tests do not match the predicted fracture model well permeabilities above a value of approximately 10^{-9} m^2 (Fig. 5). This is likely due to the pumping limitation inherent in well tests. The limitation of well tests can be examined using the Thiem equation and examining the pumping rates required to obtain a given drawdown for different values of permeability (Thiem, 1906). For a permeability of

10^{-9} m^2, relative to a distance of 1000 meters, a well with a radius of 0.25 meters would require a pumping rate of 1.26 m^3 s^{-1} (20,000 gpm) to obtain a meter of drawdown in the Edwards. This flow rate is unreasonable to use for well testing. Therefore, few standard well tests or specific capacity tests can yield values above 10^{-9} m^2 for this aquifer. Above this value, permeabilities will appear 'infinite' with these testing methods.

The fracture model requires a low number of well-connected fractures to reproduce the well-scale permeabilities. This low fracture density indicates that *a priori* predictions of well-scale permeabilities using this technique would require information on fracture connectivity, roughness, channeling, and fracture length, but that the values are feasible for the Edwards aquifer. The model also indicates that for many of the high flow wells of the Edwards, the formation is flowing under non-linear laminar ($10 < R_e < 2000$) to turbulent ($R_e > 2000$) conditions.

7.3 *Do conduits control the Edwards' permeability?*

Conduits in the Edwards aquifer do not appear to contribute significantly to the average permeability in wells (Figs 5 and 6, Model D). Conduits would produce permeabilities above the observed values for the majority of the wells. This is not unexpected because of there is a low probability of intersecting a well-connected conduit in a karst aquifer. Quinlan & Ewers (1985) estimate this at 1 in 2600 for shallow karst in Kentucky. For the Edwards which will have both conduits produced by the normal carbonic acid reaction as

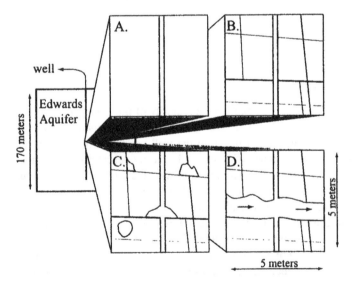

Figure 6. Four possible models of heterogeneities controlling the response of Edwards aquifer wells. Model A represents a well located in a carbonate matrix that does not intercept any fractures or conduits. Model B represents a well that intercepts fractures. Model C represents a well that intercepts fractures and dissolution features of limited lateral extent resulting in a well with an irregular wall and bit drops during drilling. Model D represents a well that intercepts a laterally continuous, hydraulically active conduit. The majority of wells in the Edwards are probably best represented by model C, but a single well may have elements of all four models. Models B, C, or D are all plausible models for causing zero drawdown wells (wells with no measurable drawdown).

in Kentucky and by mixing and/or oxidation of hydrogen sulfide, the distribution of conduits may be denser. However, it is unlikely that this probability will approach the 1 in 6 probability observed in the Edwards for the occurrence of a zero drawdown well.

The lack of continuous dendritic conduits draining the Edwards aquifer is observed at springs which discharge from fractures as opposed to conduits. The regional models also confirm this observation where regions of high permeability do not extend from the recharge to the discharge zone (Thorkildsen & McElhaney, 1992), however, these models may not have the resolution necessary to indicate conduit flow. Some research has indicated the presence of conduits associated with some springs, but it is difficult to determine the lateral extent of these conduits at depth. The presence of cave fish exiting wells, and the locations of high regional permeability do indicate that conduits control some areas of the Edwards aquifer.

7.4 *Do fractures or conduits control high permeability wells?*

A common debate among workers in the Edwards is whether the high permeabilities observed in many wells are due to fractures or conduits. This research suggests that many of the high yield wells are due to the numerous fractures that exist in the Edwards (Wermund et al., 1978). Alexander (1990) has found a correlation for this observation in the Barton Springs segment of the Edwards aquifer. It appears that the bit drops that are encountered during drilling in the Edwards are usually laterally discontinuous dissolution features which are only hydraulically connected by fractures (Fig. 6, Model C). The large amount of fracturing observed in the aquifer on many scales would support the hypothesis that the fractures are controlling the well-scale permeability and portions of the regional-scale permeability.

If conduits did contribute significantly to wells, it is unlikely that standard well tests or specific capacity tests could be used to obtain quantitative data about those wells because of minimal drawdowns (Fig. 6, Model D). The large percentage and spatial distribution of zero drawdown wells also does not support conduit flow (Fig. 4). Their pattern does not rule out a number of widely spaced conduits, but does not support the concept of a limited number of regionally significant conduits. The large percentage of zero drawdown wells, combined with the low probability of intercepting a conduit, makes it unlikely that conduits are the only contributors to high permeability wells; fractures must play a significant role.

Fractures and conduits both control flow through the Edwards aquifer. These models indicate that information on the fracture properties of the Edwards aquifer are needed to make *a priori* assessments of permeability for particular sites. The models also indicate that although dissolution features are often intercepted while drilling wells, the dissolution features are unlikely to be hydraulically continuous over long distances, and simply act to increase the effective radius of the well. In both cases, high permeability zones of the Edwards are flowing under non-Darcian conditions.

8 CONCLUSIONS

From the models used to test the scale effect for permeability in the Edwards aquifer, we conclude that:

1. Increasing permeability with scale is expected for a fractured karst aquifer. If the permeability did not increase, it would be difficult to explain how the fractures and conduits are functioning unless they were entirely unconnected.

2. Well-scale permeability in the Edwards aquifer can be modeled as predominantly a function of fracture aperture distribution. Even when no measurable drawdown occurs during pumping tests, well permeability can be explained with fractures observed in roadcuts. However, in these wells with no measurable drawdown, flow is likely non-Darcian.

3. Well-connected conduits do not appear to contribute significantly to well permeability for the average Edwards well. The probability of intersecting conduits by a large number of scattered wells is low, and fractures can provide similar responses.

4. The Edwards aquifer has some finite high permeability for which the standard well test used in the Edwards is an insufficient method to estimate permeability in approximately 15% of the cases. The standard method is also inadequate because of the high pumping rates required and discharge produced. Non-linear laminar ($10 < R_e < 2000$) to turbulent ($R_e > 2000$) flow probably occurs in these wells.

5. Turbulent flow in conduits may influence regional flow in the Edwards aquifer. This is supported by modeling and evidenced by wells that have large bit drops and produce blind catfish from wells. This turbulent flow is localized within the conduits and would not affect many well-scale and small-scale measurements.

6. Small-scale data, particularly fracture apertures, fracture density and conduit geometry, can be extrapolated to reproduce well- and regional-scale permeabilities. However, without sufficient connectivity, length, roughness, and channeling information, this is difficult to do *a priori*.

ACKNOWLEDGMENTS

We thank Sue Hovorka of the Texas Bureau of Economic Geology for providing access to her matrix permeability data. We would also like to thank Art Palmer and George Veni for their thoughtful and thorough reviews of this paper. Funding for Todd Halihan was provided by a National Science Foundation Traineeship in Hydrology (NSF grant GER-9454098). Funding for permeability data collection was provided by the Edwards Aquifer Authority. Contact R. Mace at the Texas Bureau of Economic Geology for a copy of the dataset. Manuscript preparation was supported by the Owen-Coates Fund of the Geology Foundation of The University of Texas at Austin.

REFERENCES

Alexander, K.B. 1990. Correlation of structural lineaments and fracture traces to water-well yields in the Edwards aquifer, central Texas (Masters Thesis). Austin: The University of Texas at Austin, 113 p.

Atkinson, T.C. 1985. Present and future directions in karst hydrogeology: *Annales. Societe Geologique de Belgique*, 108: 293-296.

Brace, W.F. 1984. Permeability of crystalline rocks: New in situ measurements. *Journal of Geophysical Research*, 89(B6): 4327.

Bradbury, K.R. & Muldoon, M.A. 1990. Hydraulic conductivity determinations in unlithified glacial and fluvial materials, in Nielsen, D.M. & Johnson, A.I. (eds) *Ground water and vadose zone monitoring*, ASTM STP 1053. Philadelphia, American Society for Testing and Materials, p. 138-151.

Clauser, C. 1992. Permeability of crystalline rocks. EOS, Transactions American *Geophysical Union*, 73(21): 233-238.

Dagan, G. 1986. Statistical theory of groundwater flow and transport: Pore to laboratory, laboratory to formation, and formation to regional scale. *Water Resources Research*, 22(9): 120S-134S.

Dyke, C. 1995. The detection and characterization of natural fracture permeability whilst drilling, in Myer, L.R., Cook, N.G.W., Goodman, R.E. & Tsang, Chin-Fu (eds) *Fractured and Jointed Rock Masses*. Rotterdam: Balkema, p. 591-601.

Fetter, C.W. 1994. *Applied Hydrogeology*. New York: Macmillan, 691 p.

Ford, D.C. & Williams, P.W. 1989. *Karst Geomorphology and Hydrology*. London: Unwin Hyman, 601 p.

Freeze, R.A. & Cherry, J.A. 1979. *Groundwater*. Englewood Cliffs, New Jersey: Prentice Hall, 604 p.

Gupta, R.S. 1989. *Hydrology and Hydraulic Systems*. Englewood Cliffs, New Jersey: Prentice Hall, 739 p.

Halihan, T., Wicks, C.M. & Engeln, J.F. 1998. Physical response of a karst drainage basin to flood pulses: Example of the Devil's Icebox cave system (Missouri, USA). *Journal of Hydrology*, 204: 24-36.

Hovorka, S.D., Mace, R.E. & Collins, E.W. 1995. Regional distribution of permeability in the Edwards aquifer, Report 95-02. San Antonio, Texas, Edwards Underground Water District, 128 p.

Hovorka, S.D., Mace, R. E. & Collins, E.W. 1998. Permeability structure of the Edwards aquifer, south Texas-Implications for aquifer management: Austin, Texas, Bureau of Economic Geology, University of Texas at Austin, Report of Investigations No. 250, 55 p.

Hovorka, S.D., Ruppel, S.C., Dutton, A.R. & Yeh, J. 1993. Edwards aquifer storage assessment, Kinney County to Hays County, Texas: Austin, Texas, Bureau of Economic Geology, University of Texas at Austin, report prepared for the Edwards Underground Water District, 101 p.

Hovorka, S.D. 1997. personal communication.

Jensen, J.L., Lake, L.W., Corbett, P.W.M. & Goggin, D.J. 1997. *Statistics for Petroleum Engineers and Geoscientists*. Upper Saddle River, New Jersey: Prentice Hall, 390 p.

Kiraly, L. 1975. Rapport sur l'état actuel des connaissances dans le domaine des caractères physiques des roches karstiques, in Burger, A. & Dubertret, L. (eds) *Hydrogeology of karstic terrains*. Paris, International Association of Hydrogeologists, Series B, No. 3, p. 53-67.

Klemt, W.B., Knowles, T.R., Elder G.R. & Sieh, T.W. 1979. Ground-water resources and model applications for the Edwards (Balcones fault zone) aquifer in the San Antonio region, Texas: Austin, Texas, Texas Water Development Board, Report 239, 88 p.

Lamb 1932. *Hydrodynamics*, 6th edition. New York: Dover, 738 p.

Leonards, G.A. 1962. *Foundation Engineering*. New York: McGraw-Hill, 1136 p.

Lindquist, E. 1933. *On the flow of water through porous soil*. Stockholm, Premier Congres des grands barrages, p. 81-101.

Long, J.C.S., Remer, J.S., Wilson, C.R. & Witherspoon, P.A. 1982. Porous media equivalents for networks of discontinuous fractures. *Water Resources Research*, 18: 645-658.

Longley, G. 1981. The Edwards aquifer: Earth's most diverse ground-water ecosystem. *International Journal of Speleology*, 11: 123-128.

Mace, R.E. 1997. Determination of transmissivity from specific capacity tests in a karst aquifer. *Ground Water*, 35(5): 738-742.

Mace, R.E. 1998. Regional distribution, spatial correlation, and scaling of permeability in a karstic aquifer, Edwards Group, south-central Texas. Ground Water, (in press).

Maclay, R.W. & Land, L.F. 1988. Simulation of Flow in the Edwards Aquifer, San Antonio Region, Texas, and Refinement of Storage and Flow Concepts. *U.S. Geological Survey Water-Supply Paper 2336-A*, 48 p.

Maclay, R.W. & Small, T.A. 1986. Carbonate geology and hydrology of the Edwards aquifer in the San Antonio area, Texas: Austin, Texas, Texas Water Development Board, Report 296, 121 p.

Marrett, R. 1996. Aggregate properties of fracture populations. *Journal of Structural Geology*, 18(2/3): 169-178.

McKinney, D.C. & Watkins, D.W., Jr. 1993. Management of the Edwards Aquifer: A Critical Assessment: Austin, Texas, Center for Research in Water Resources, Technical Report CRWR 244, 94 p.

Neuzil, C.E. 1994. How permeable are clays and shales? *Water Resources Research*, 30(2): 145-150.

Odling, N.E. 1997. Scaling and connectivity of joint systems in sandstones from western Norway. *Journal of Structural Geology*, 19(10): 1257-1271.

Pavlicek, D.L., Small, T.A. & Rettman, P.L. 1987. Hydrogeologic data from a study of the freshwater zone/saline water zone interface in the Edwards aquifer, San Antonio region, Texas: U.S. Geological Survey Open-File Report 87-389, 108 p.

Poteet, D., Collier, H. & Maclay, R. 1992. Investigation of the fresh/saline-water interface in the Edwards aquifer in New Braunfels and San Marcos, Texas: San Antonio, Texas, Edwards Underground Water District, Report 92-02, 176 p.

Quinlan, J.F. & Ewers, R.O. 1985. Ground water flow in limestone terranes: strategy rationale and procedure for reliable, efficient monitoring of ground water quality in karst areas, in *Proceedings of the Fifth National Symposium and Exposition on Aquifer Restoration and Ground Water Monitoring*. Worthington, Ohio, National Water Well Association, p. 197-234.

Quinlan, J.F., Davies, G.J. & Worthington, S.R.H. 1992. Rationale for the design of cost-effective groundwater monitoring systems in limestone and dolomite terranes: Cost-effective as conceived is not cost-effective as built if the system design and sampling frequency inadequately consider site hydrogeology, in *Symposium on Waste Testing and Quality Assurance (8th, Washington, D.C., July 1992)*. Washington, D.C., US Environmental Protection Agency, p. 552-570.

Rasband, W. 1994. *NIH Image, Version 1.50*. U.S. National Institutes of Health.

Rehfeldt, K.R., Hufschmied, P., Gelhar, L.W. & Schaefer, M.E. 1989. Measuring hydraulic conductivity with the borehole flowmeter. Palo Alto, California, Electric Power Research Institute, Research Report EPRI EN-6511, Project 2485-5, 209 p.

Rose, P.R. 1972. *Edwards group, surface and subsurface, central Texas*. Austin, Texas, University of Texas, Bureau of Economic Geology Report of Investigations 74, 198 p.

Rovey, C.W., II. 1994. Assessing flow systems in carbonate aquifers using scale effects in hydraulic conductivity. *Environmental Geology*, 24: 244-253.

Scheidegger, A.E. 1974. *The physics of flow through porous media, 3rd edition*. Toronto, University of Toronto Press, 353 p.

Sharp, J.M., Jr. & Banner, J.L. 1997. The Edwards aquifer: A resource in conflict. GSA *Today*, 7(8): 1-9.

Streeter, V.L. 1948. *Fluid Dynamics*. New York: McGraw-Hill, 480 p.

Swanson, G.J. 1991. Super well is deep in the heart of Texas. *Water Well Journal*, no. 7, p. 56-58.

Thiem, G. 1906. *Hydrologische methoden*. Leipzig, Gebhardt, p. 56.

Thorkildsen, D. & McElhaney, P.D. 1992. Model refinement and applications for the Edwards (Balcones fault zone) aquifer in the San Antonio region, Texas: Austin, Texas, Texas Water Development Board, Report 340, 33 p.

Thornhill, P.D., Harden, R.W. & Nevola, R. 1988. The Edwards aquifer, underground river of Texas: Seguin, Texas, Guadalupe-Blanco River Authority Publication, 63 p.

Turcotte, D.L. & Schubert, G. 1982. *Geodynamics: Applications of continuum physics to geological problems*. New York, Wiley, 450 p.

Veni, G. 1988. *The Caves of Bexar County, 2nd edition:* Austin, Texas Memorial Museum, The University of Texas at Austin, 300 p.

Wermund, E.G., Cepeda, J.C. & Luttrell, P.E. 1978. Regional distribution of fractures in the southern edwards plateau and their relationship to tectonics and caves. Austin, Texas, *Bureau of Economic Geology, Geological Circular 78-2*, The University of Texas at Austin, 14 p.

A numerical approach to simulating mixed flow in karst aquifers

JOHN J. QUINN & DAVID TOMASKO
Environmental Assessment Division, Argonne National Laboratory, Argonne, IL, USA

ABSTRACT: Modeling groundwater flow in a karst environment is both numerically challenging and highly uncertain because of potentially complex flowpaths and a lack of site-specific information. This study presents the results of numerical modeling in which drain cells in a finite-difference model are used as analogs for preferential flowpaths or conduits in karst environments. In this study, conduits in mixed-flow systems are simulated by assigning sequences of adjacent drain cells from the locations of tracer releases, sinkholes, or other karst features to outlet springs along inferred flowpaths. These paths are determined by the locations of losing stream segments, ephemeral stream beds, fracture lineaments, other surficial characteristics, or results of geophysical surveys, combined with the results of dye traces. The elevations of the drains at the discharge ends of the inferred flowpaths are set to the elevations of discharge springs; the elevations at the beginning of the inferred flowpaths are estimated from field data and are adjusted as necessary during model calibration. To simulate a free-flowing conduit, a high conductance is assigned to each drain to promote the removal of water by modeled conduits and to eliminate the need for drain-specific information that would be very difficult to obtain. Calculations were performed for two sites: one near Hohenfels, Germany, and one near St. Louis, Missouri. The potentiometric surfaces produced by these simulations agreed well with field data. The head contours in the vicinity of the karst features behaved in a manner consistent with a flow system having both diffuse and conduit components, and the sum of the volumetric flow out of the drain cells agreed closely with spring discharges and stream flows. Because of the success of this approach, it is recommended for testing of conceptual flow models and for regional studies in which little site-specific information is available, and general flow characteristics are desired.

1 INTRODUCTION

Conduit flow will prevail or at least be present in carbonate terrains in which any of the following features are present: sinkholes, dry valleys, sinking streams, or karren (Rovey, 1994). The aquifers with the highest proportion of non-Darcian flow are probably those composed of limestone, where fissures and dissolution features may be enlarged to form conduits that can range in size from 0.01 to 10 m in diameter (Gale, 1983).

As discussed in the literature, modeling groundwater flow in a karst environment can be challenging and often produces results that are highly uncertain because of the complexity of flowpaths and sparsity of site-specific information. In the past, many modeling approaches have been used to simulate flow in a karst environment: models using an equivalent porous medium in which flow is governed by Darcy's law (Anderson & Woessner, 1992); 'black-box' approaches in which functions are developed to reproduce input and output system responses (recharge and flow at discharge springs) (e.g., Dreiss, 1989a,b); models in which the preferred flowpath is simulated with a very high hydraulic conductivity relative to the surrounding matrix material (up to eight orders of magnitude difference) (e.g., Teutsch, 1989; Mace, 1995; Eisenlohr et al., 1997); fracture network simulations in which individual fractures are mapped and then studied (Long et al., 1982; Long & Billaux, 1987); and open channel equivalents (Thrailkill et al., 1991).

A mixed-flow system in a karst environment is a combination of diffuse and conduit flow (Quinlan & Ewers, 1985; Field, 1993). Simulation of a karst system composed of dendritic paths (Milanovic, 1981; White, 1988; White & White, 1989) may require a great deal of site-specific information for the karst channels and flow conditions (e.g., elevation, slope, fill material, roughness, cross-sectional area, Reynolds number, Froude number, diameter, etc.) (Field & Nash, 1997; Field, 1997). Because this information is difficult, if not impossible, to obtain, flow modeling in karst terrains is generally not performed, or simplifying assumptions are used.

For this study, preferred flowpaths in a karst environment are simulated by using the drain feature of the finite-difference code MODFLOW (McDonald & Harbaugh, 1988). The beginning of the preferred flowpath coincides with the location of a known surficial feature (e.g., sinkhole, lineament, fracture), especially those that were the location of a dye release, and terminates at a surficial discharge point. Intermediate points are assigned on the basis of an inferred flowpath determined from losing stream segments, ephemeral stream beds, fracture lineaments, other surficial characteristics, or results of geophysical surveys combined with the results of dye traces. The total discharge for the modeled conduit is the sum of the discharges of each drain along the flowpath.

This approach is similar to using a discrete singular fracture set model (Teutsch & Sauter, 1991) without incorporating detailed information on the fractures. Rather, the modeling addresses key hydrogeologic features on a scale of less than 100 m to several kilometers. This scale is most important when considering flow and transport (Thrailkill, 1986). By using this method, the numerical instability associated with modeling an extremely high permeability contrast between a preferential flowpath and the adjacent aquifer materials is avoided.

2 MODFLOW DRAIN

The drain feature in MODFLOW was developed to simulate agricultural tiles that remove water from an aquifer at a rate proportional to the difference in water level (head) between the aquifer and some fixed drain elevation, as long as the head in the aquifer is above that elevation (McDonald & Harbaugh, 1988). If the head in the aquifer falls below that of the drain, no additional water removal occurs. For the computations presented in this study, drain elevations at the discharge points of the preferred flowpaths were assumed to be equal to the elevations of associated springs or levels in surficial receiving waters. At the upstream end of the flowpaths, the elevations were estimated from drilling logs, potentiometric maps of the shallow groundwater systems, and bedrock maps. Elevations of drains in model cells located along the flowpath were estimated by linear interpolation between the two ends to produce smooth transitions between cells.

Adjusting drain elevations is a means of achieving model calibration. By changing the drain elevations and the length of inferred conduits, the match to target heads at site monitoring wells can be improved.

In addition to drain elevations, the drain conductance must also be specified. This lumped parameter incorporates information on characteristics of the drain and its immediate surroundings, as well as the head loss between the drain and the aquifer (McDonald & Harbaugh, 1988). For simplicity, a high conductance value was selected to eliminate the need for drain-specific data that would be difficult to obtain. A value of 100 m/d per meter of conduit length was converted to drain conductance for each drain cell. This high value of conductance promotes the removal of water from the simulated conduits. Although uncertainty is associated with the conductance assigned to the drains, use of this high value produces a reasonable effect on potentiometric surfaces.

Prior applications of MODFLOW in karst settings have made use of the drain package or the general head boundary package by assigning a drain (e.g. Yobbi, 1989) or a general head boundary (e.g. Dufresne & Drake, 1999) to each model cell representing an outlet spring. However, the flow removed from the system by the drain or general head boundary is limited to seepage from the adjacent upgradient model cells; the contributions from upgradient conduits are ignored. In the current study, connected pathways of drain cells simulate the conduit portion of mixed-flow karst aquifers and are conceptually more accurate and realistic.

3 CASE STUDIES

Two case studies were performed to evaluate the potential for using MODFLOW drain cells to simulate flow in a karst environment. The first was performed for the Lautertal dry valley, near Hohenfels, Germany. The second was done for a site in Missouri.

3.1 *Lautertal valley, Hohenfels, Germany*

The Lautertal dry valley is part of the Combat Maneuver Training Center (CMTC) near Hohenfels, Germany (Fig. 1). In the vicinity of the valley, the shallow groundwater system is complex. Precipitation is subject to rapid runoff and infiltration through solution-widened fractures. No perennial surface water features exist; however, local ponding occurs in the vicinity of dolines. Some of the infiltrating precipitation discharges rapidly to springs in the Lauterach River valley (Glennon et al., unpublished information, 1998). The remainder of the water recharges the near-surface aquifer. This aquifer consists of a network of joints, fractures, and cavities in the carbonate Malm Formation (Heigold et al., 1997). The overall direction of groundwater flow is toward the Vils River.

On the basis of geophysical analyses (Heigold et al., 1997; Thompson et al., unpublished information, 1997a,b), the shallow groundwater system consists of the weathered

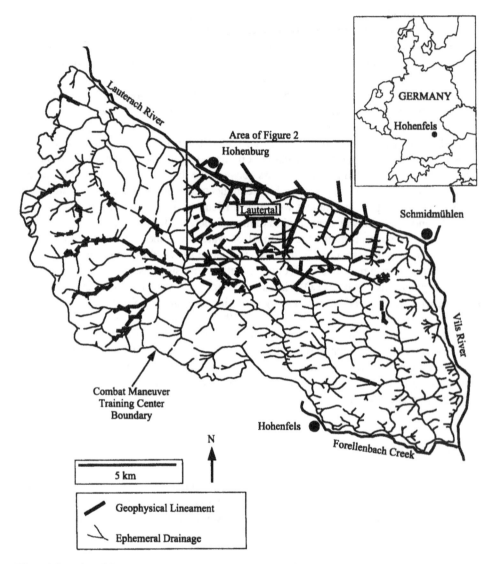

Figure 1. Location of the Lautertal study area, geophysical lineament data, and on-base ephemeral drainages.

and fractured portion of the upper Malm Formation. Recharge to the shallow flow system occurs by infiltration of precipitation and interaction with dolines, losing stream segments, and fractures.

The Lautertal modeling domain includes the entire CMTC and was extended off base to the west and south to make use of regional flow boundary conditions (Quinn et al., 1998). Figure 1 shows the CMTC portion of the 10 by 30 km modeling area. More than 1,800 drain cells are concentrated in the portion of the modeling domain that has abundant geophysical lineament results and several tracer tests. Because the Lautertal area is the focus of this study and has the highest density of geophysical results, the grid in this area has the highest resolution, with cells that are 100 by 100 m. Near the model boundaries, the grid-cell density decreases, and the grid-cell size is 500 by 500 m.

The Hohenfels site has a limited monitoring well network – six wells located near the center of the modeling domain – to be used for calibration purposes. A suitable calibration of the model was achieved by adjusting drain elevations to ensure active flow in all inferred conduits, then adjusting hydraulic conductivity to match the heads at the well locations (Glennon et al., unpublished information, 1998).

Figure 2 illustrates the implementation of the drains in conjunction with geophysical lineament data in the Lautertal vicinity, together with the flow vectors of the calibrated model. Localized changes in the direction of groundwater flow illustrate the interaction of groundwater with the adjacent surface water bodies and suggest that the boundary conditions for the model are defensible. That is, at the Forellenbach, Vils, and Lauterach rivers, water-level contours bend upstream in an expected fashion. At the modeled conduits, the contours behave in the manner expected for a mixed-flow karst environment: a

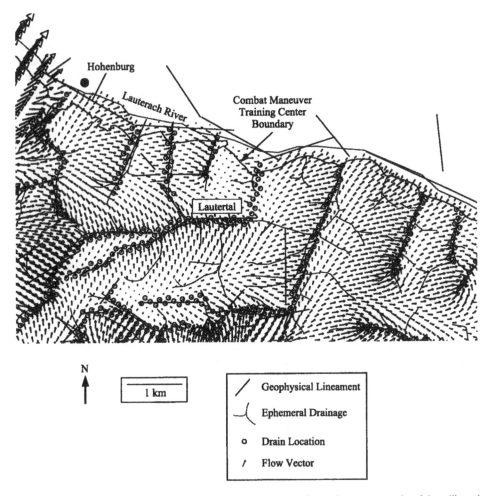

Figure 2. Drain distribution in the vicinity of the Lautertal and groundwater flow vector results of the calibrated model.

combination of diffuse and conduit flow. Groundwater converges on the preferential flowpath, and lines of equal potential point upstream.

The calibrated model's water budget indicates that, in the portion of the modeling area containing drains, most of the water that enters the system as recharge exits as conduit flow to springs rather than as diffuse discharge to the northern boundary, the Lauterach River. The code ZONEBUDGET (Harbaugh, 1990) permits a detailed analysis of the MODFLOW water budget. The predicted groundwater discharge along the southern bank of the Lauterach River between the westernmost and easternmost drains (approximately two-thirds of the model's width along the Lauterach) indicated a steady-state diffuse discharge of 46 L/s and a total conduit discharge of 810 L/s. The summed total discharge (856 L/s) is 30% of the mean measured river flow in the Lauterach River at Schmidmühlen (Fig. 1) of 2800 L/s (Rothascher, 1987). Considering that approximately the same amount of flow enters the Lauterach from north of the modeled area, and additional water enters the river as groundwater from the remaining upstream third of both banks and as runoff, the results indicate reasonable agreement between the steady-state simulation and the mean field data.

In addition, one-time spring gaging measurements are available for springs at Schmidmühlen (73 L/s) and Papiermühle/Hohenburg (205 L/s) (Herausgegeben vom Bayerischen Landesamt für Wasserwirtschaft München, 1990). Summing the flows from these major springs and others along the detailed portion of the Lauterach's southern bank could give a total flow similar to the 810 L/s calculated by the model. The calibration of the model could be improved if more spring gaging data were available.

3.2 *Missouri site*

The second case study focuses on a site in east-central Missouri (Fig. 3) where the Burlington-Keokuk Formation crops out (Rovey, 1994). This formation is particularly susceptible to karstification, and wherever it is exposed in Missouri, there are karst features. Elsewhere in the state, this unit has been determined to exhibit a combination of conduit flow and diffuse flow (Wicks, 1997). Numerous dye traces performed at and near the site by the Missouri Department of Natural Resources (MDNR) (MDNR, 1991) confirmed the presence of conduit flow, with recorded velocities of approximately 1000 m/d, assuming straight travel paths. Tracers were released in swallows, losing streams, and boreholes. The tracing also indicated that several preferential pathways north of the site jump across drainages. Other hydraulic data are consistent with the presence of conduit flow. These data include drill-bit drops of up to 1.5 m during monitoring well installation, the presence of pronounced troughs in the potentiometric surface of the shallow groundwater aquifer, and the presence of lows in the bedrock topography along inferred paleochannels. The site is located on a groundwater and surface water divide. Water north of the site flows to a Mississippi River tributary; water south of the site flows directly to the Missouri River.

As for the Lautertal dry valley study, a MODFLOW model was developed to predict groundwater flow within the karst environment. The modeling domain was extended from the site to regional hydrologic boundaries. The central portion of the modeling domain is shown in Figure 3. For this model, more than 600 cells incorporated drains along inferred conduits. Much of the conduit system north of the site appears to be convergent, with discharge occurring at a spring roughly 2 km to the north.

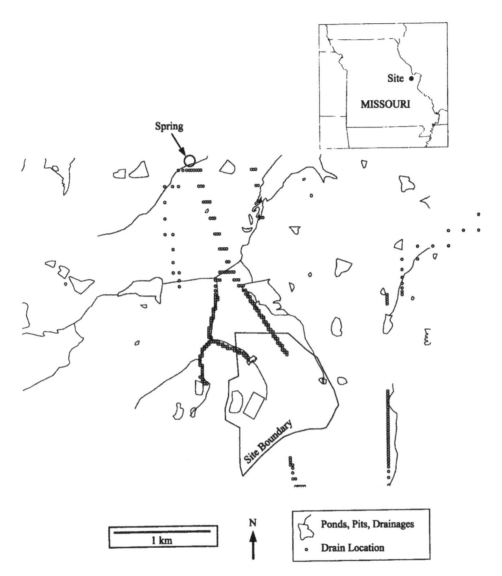

Figure 3. Location of Missouri site and drain distribution in the center of the modeling area.

An important aspect of this site in assessing the accuracy of this modeling approach is its extensive monitoring well network. The spatial head distribution was useful in calibrating the model by adjusting the drain elevations and the hydraulic conductivity (Quinn et al., unpublished data, 1997).

ZONEBUDGET (Harbaugh, 1990) was again used to analyze the MODFLOW water balance for the calibrated potentiometric surface (Fig. 4). The comparison of the simulated steady-state discharge at the spring with field-measured flows was excellent; the sum of the water removed from the associated drain cells was within 14% of the annual average spring discharge (U.S. Geological Survey, 1995).

154   *John J. Quinn & David Tomasko*

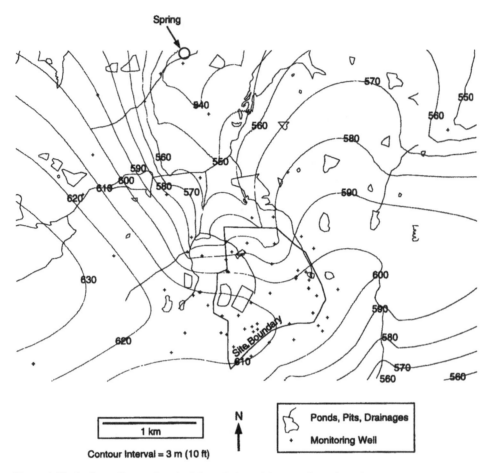

Figure 4. Monitoring well network at the Missouri site and the potentiometric surface results of the calibrated model.

4 SUMMARY

This paper presented a method of numerically modeling mixed flow in a karst environment by using sequences of adjacent model cells with drains to simulate conduits. The discharge ends of the preferred flowpaths were chosen to coincide with springs or other surface water features; initial points coincided with surficial features, such as dolines, geophysically determined lineaments, or the release locations of successful dye traces. Drain cells between the end points were assigned along the inferred conduit system. At the discharge end of the preferential flowpath, drain elevations were assigned on the basis of elevations of discharge springs and receiving waters. At the inlet end, the elevations were estimated from drilling records, potentiometric surface and bedrock elevation maps, and were then adjusted during calibration. Intermediate elevations were estimated by linear interpolation. Drain conductance was set high to promote removal of groundwater by modeled conduits and to eliminate the need for site-specific information. Total discharge from each conduit was the sum of the discharge from each drain along the conduit's flowpath.

Comparisons between the results obtained with the MODFLOW drain models and field measurements (potentiometric surface and volumetric discharge) ranged from good at a site near Hohenfels, Germany, to excellent at a site in Missouri. Because of the success of this method, it is recommended for testing of conceptual hydrogeologic models and for regional studies with limited site-specific information (e.g., location, number, size, and conductivity of fractures and conduits).

ACKNOWLEDGMENTS

This work was supported in part by the U.S. Department of Defense, U.S. Army, under interagency agreement, through U.S. Department of Energy contract W-31-109-Eng-38 and in part by the U.S. Department of Energy, Office of Environmental Restoration and Waste Management, under contract W-31-109-Eng-38. The authors extend thanks to Ward Sanford and Malcolm Field for their helpful reviews.

REFERENCES

Anderson, M.P. & Woessner, W.W. 1992. *Applied groundwater modeling: simulation of flow and advective transport.* New York: Academic Press, Inc., 381 p.

Dreiss, S.J. 1989a. Regional scale transport in a karst aquifer. 1. component separation of spring flow hydrographs. *Water Resources Research,* 25(1): 117-125.

Dreiss, S.J. 1989b. Regional scale transport in a karst aquifer. 2. Linear systems and time moment analysis. *Water Resources Research,* 25(1): 126-134.

Dufresne, D.P. & Drake, C.W. Regional groundwater flow model construction and wellfield site selection in a karst area, Lake City, Florida. *Engineering Geology,* 52: 129-139.

Eisenlohr, L., Bouzelboudjen, M., Kiraly, L. & Rossier, Y. 1997. Numerical versus statistical modeling of natural response of a karst hydrogeological system. *Journal of Hydrology,* 202: 244-262.

Field, M.S. 1993. Karst hydrology and chemical contamination. *Journal of Environmental Systems,* 22(1): 1-26.

Field, M.S. 1997. Risk assessment methodology for karst aquifers: (2) solute-transport modeling. *Environmental Monitoring and Assessment,* 47: 23-37.

Field, M.S. & S.G. Nash 1997. Risk assessment methodology for karst aquifers: (1) estimating karst conduit-flow parameters. *Environmental Monitoring and Assessment,* 47: 1-21.

Gale, S.J. 1983. The hydraulics of conduit flow in carbonate aquifers. *Journal of Hydrology,* 70: 309-327.

Harbaugh, A.W. 1990. A computer program for calculating subregional water budgets using results from the U.S. Geological Survey modular three-dimensional finite-difference ground-water flow model. U.S. Geological Survey, Open-File Report 90-392, 24 p.

Heigold, P.D., Thompson, M.D. & Borden, H.M. 1997. Geophysical exploration in the Lautertal at the Combat Maneuver Training Center, Hohenfels, Germany: Argonne, Illinois, Argonne National Laboratory, ANL/ESD/TM-82.

Herausgegeben vom Bayerischen Landesamt für Wasserwirtschaft München, 1990, Verzeichnis der Quellen in Bayern: München, Bayerischen Landesamt für Wasserwirtschaft.

Long, J.C.S. & Billaux, D.M. 1987. From field data to fracture network modeling: an example incorporating spatial structure. *Water Resources Research,* 23(7): 1201-1216.

Long, J.C.S., Remer, J.S., Wilson, C.R. & Witherspoon, P.A. 1982. Porous media equivalents for networks of discontinuous fractures. *Water Resources Research,* 18(3): 645-658.

Mace, R.E. 1995. Geostatistical description of hydraulic properties in karst aquifers: a case study in the Edwards Aquifer, in *Proceedings, International Symposium on Groundwater Management.* New York, American Society of Civil Engineers, p. 193-198.

McDonald, M.G. & Harbaugh, A.W. 1988. A modular three-dimensional finite-difference ground-water flow model. *Techniques of Water-Resources Investigations*, Book 6, Chapter A1, U.S. Geological Survey.

Milanovic, P.T. 1981. *Karst hydrogeology*. Littleton, Colorado, Water Resources Publications.

Missouri Department of Natural Resources, 1991, Shallow groundwater investigations at Weldon Spring, Missouri: final report for fiscal years 1988-1990: Rolla, Missouri, Division of Geology and Land Survey, prepared for the U.S. Department of Energy, Weldon Spring Site Remedial Action Project, St. Charles, Missouri, June.

Quinlan, J.F. & Ewers, R.O. 1985. Ground water flow in limestone terraces: strategy, rationale and procedure for reliable, efficient monitoring of ground water quality in karst areas, in *Proceedings, Fifth National Symposium on Aquifer Restoration and Ground Water Monitoring*, Dublin, Ohio, National Water Well Association, p. 197-234.

Quinn, J., Tomasko, D., Glennon, M.A., Miller, S.F. & McGinnis, L.D. 1998. Using MODFLOW drains to simulate groundwater flow in a karst environment, in *Proceedings of MODFLOW '98*. Golden, CO, International Ground Water Modeling Center, p. 105-112.

Rothascher, A. 1987. Die Grundwasserneubildung in Bayern: München, Informationsberichte Bayerisches Landesamt für Wasserwirtschaft.

Rovey, C.W. 1994. Assessing flow systems in carbonate aquifers using scale effects in hydraulic conductivity. *Environmental Geology*, 24: 244-253.

Teutsch, G. 1989. Groundwater models in karstified terrains: two practical examples from the Swabian Alb (S. Germany), in *Proceedings, Solving Ground Water Problems with Models*. Indianapolis, Indiana, International Ground Water Modeling Center, p. 929-953.

Teutsch, G. & Sauter, M. 1991. Groundwater modeling in karst terrains: scale effects, data acquisition, and field validation, in *Proceedings, Third Conference on Hydrogeology, Ecology, Monitoring, and Management of Ground Water in Karst Terrain*. Dublin, Ohio, National Water Well Association, p. 17-35.

Thrailkill, J. 1986. Models and methods for shallow conduit-flow carbonate aquifers in *Proceedings, Environmental Problems in Karst Terranes and Their Solutions Conference*. Dublin, Ohio, National Water Well Association, p. 17-31.

Thrailkill, J., Sullivan, S.B. & Gouzie, D.R. 1991. Flow parameters in a shallow conduit-flow carbonate aquifer, Inner Bluegrass Karst Region, Kentucky, USA. *Journal of Hydrology*, 129: 87-108.

U.S. Geological Survey 1995. *Weldon Spring Quarry ground-water level monitoring program, third quarter 1995*. Rolla, Missouri, U.S. Geological Survey, October.

White, W.B. 1988. *Geomorphology and hydrology of karst terrains*. New York, Oxford University Press, 464 p.

White, W.B. & White, E.L. 1989. *Karst hydrology concepts from the Mammoth Cave Area*. New York, Van Nostrand Reinhold, 346 p.

Wicks, C.M. 1997. Origins of groundwater in a fluviokarst basin: Bonne Femme basin in central Missouri, USA. *Hydrogeology Journal*, 5(3) 89-96.

Yobbi, D. 1989. Simulation of steady-state ground water and spring flow in the upper Floridian aquifer of coastal Citrus and Hernando Counties, Florida: Tallahassee, Florida, U.S. Geological Survey, Water-Resources Investigations Report 88-4036, 33 p.

Studies of confluent flow in mature karst aquifers using analog models and numerical mixing cell models

WANFANG ZHOU
P.E. LaMoreaux & Associates, Inc., Oak Ridge, Tinnessee, USA

ABSTRACT: As a prolific source of groundwater and a geologic medium vulnerable to contamination, karst aquifers have been studied to a great extent in many countries. However, comprehensive theoretical models for mature karst aquifers have not fully developed yet. Three analog models were constructed to study the confluent flow properties in karst aquifers. The experimental results revealed a non-linear relationship between water flow velocity and hydraulic gradient. Under circumstances where the Darcy's law is no longer valid, finite-difference mixing cell models were proposed to simulate groundwater flow and solute transport in mature karst aquifers. They assume only the mass conservation and two mixing rules in each cell. A dye tracer test was interpreted using the developed model. By adjusting the number of cells, the water volume of each cell, and the time step, a reasonable fit was achieved. The interpreted model consists of six cells with water volume of 75 m^3 in each. The design of the finite-difference mixing cell model and the initial guess of flow circulation require some a priori knowledge of geologic and hydrologic conditions.

1 INTRODUCTION

As a prolific source of groundwater and a geologic medium vulnerable to contamination, karst aquifers have been studied to a great extent in many countries. In order to develop a more consistent understanding of karst terranes and their variability, various attempts have been made employing classification schemes which are based on descriptions of the various flow phenomena in the fields (White, 1977; Smart & Hobbs, 1986). Although very useful in providing a conceptual framework for the description of various observations made in karst areas, there still remains a need for a more quantitative analysis of the field data commonly available. In contrast, flow in karst aquifers is still considered hardly quantifiable. Predictions based on conventional investigation methods seem to fail quite frequently, similarly most attempts to employ standard mathematical models did not prove satisfactory so far (Teutsch & Sauter, 1991).

The difficulty in analyzing groundwater flow or contaminant transport in karst aquifers lies in their complicated flow regimes in which the traditional Darcy's law describing groundwater flow through porous media is challenged. Darcy's law was derived from experiments passing water through a vessel filled with saturated sands and it describes a linear relationship between water flow velocity and hydraulic gradient. As pointed out by Freeze & Cherry (1979), Darcy's law is only valid when flow is laminar. Considering the wide range of water-bearing media such as conduits, fractures and pores with various sizes in mature karst aquifers, one may speculate the existence of turbulent flow under certain circumstances (Zhou, 1990). Turbulent flow in individual conduits has been studied by many researchers (Ewers, 1978) and has been described by Chezy's law where the velocity is directly proportional to the square root of the hydraulic gradient.

However, such regular relationships between groundwater flow velocity and hydraulic gradient can be rarely seen in mature karst aquifers. Figure 1 shows the dye tracer results at Zhaixiangkou dam test site of south China. There is no obvious correlation between the average flow velocity and the hydraulic gradient. Instead, statistical analyses led to the following equation:

$$v = h / (0.013 - 0.032 / I) \quad r = -0.975 \tag{1}$$

where v is the linear groundwater flow velocity derived from tracer tests, h is the potential difference between tracer injection and detection points, I is the hydraulic gradient, and r is the correlation coefficient.

By statistically analyzing tracer tests in other karst areas, Zou (1994) has suggested the following three relationships between v and I:

$$v = h / (a + b / I) \tag{2}$$

$$v = h * a * e^{bI} \tag{3}$$

$$v = h * (A_0 + A_1 I + A_2 I^2 + A_3 I^3) \tag{4}$$

where a, b, A_0, A_1, A_2, and A_3 are regression constants.

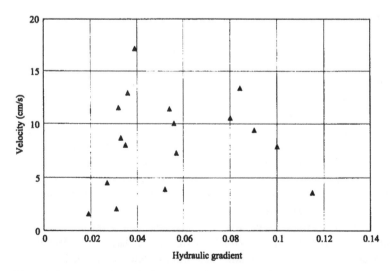

Figure 1. Groundwater flow velocity versus hydraulic gradient from dye tracer tests at jaixiangkou dam of china.

Equations (1) through (4) indicate that groundwater flow velocity in mature karst aquifers is not only related to the hydraulic gradient but also to the potential difference. Because each relationship was established from tracer tests conducted in the same aquifer with different injection and detection points, the data are an indication of heterogeneity of the aquifer. A similar conclusion has also been derived from the groundwater tracing results in the Dinaric karst of Yugoslavia (Milanovic, 1981). Because mature karst aquifers are often characterized by strong anisotropy and heterogeneity (Ford & Williams, 1989), neither Darcy's law or Chezy's law is readily applicable in describing groundwater flow in them.

The characteristic features of mature karst aquifers are the conduits that act as master drains for the groundwater system. Figure 2 illustrates a conceptual model of groundwater flow in mature karst aquifers and compares the different flow characteristics between a mature karst aquifer and a granular aquifer. The active conduits receive infeeders from the sinkhole drains and there are upstream infeeders where the sinking streams connect. Because of their minimal hydraulic resistance, conduit systems will not support much hydraulic head and form a groundwater trough. In other parts of the aquifer on both sides of the conduit, however, the groundwater level may stand much higher and thus flow toward the conduit. As a result, distributary patterns are developed so that the discharge from the whole aquifer occurs at discrete springs. The developed drainage pattern at the scale of a borehole appears random, but at the scale of massif appears organized. This organization is generally analogous to channel patterns on river deltas where rivers may have many mouths. We call this phenomenon of channelization and concentration of flow along discrete conduits 'confluent flow'. The different characteristics of the confluent flow in mature karst aquifers as compared to a granular aquifer means that groundwater

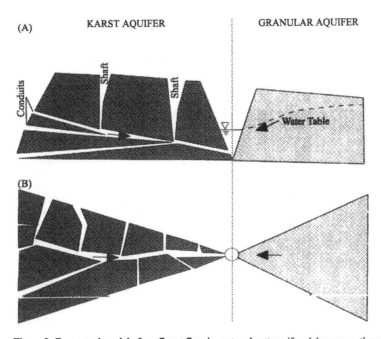

Figure 2. Conceptual model of confluent flow in mature karst aquifers (a) cross-section view; (b) plain view.

flow lines cannot be reliably intersected with a randomly drilled monitor well (Quinlan, 1990).

2 ANALOG MODELS

Figure 3 shows three experimental structures used for simulating the confluent flow in karst aquifers. The recharge was controlled by upstream water tanks. The karst conduits were represented by plastic pipes with different diameters. The discharge was either free or controlled by a downstream water tank. The first model was designed to simulate the hydrodynamic properties of a karst aquifer with two connected conduits at different levels. When water level was lower than the elevation of the higher pipe, water flows through only the lower pipe. The second one is to study the effect of two connected horizontal conduits on groundwater flow. The third one is similar to the first one but has three conduits and a downstream water tank confines the discharge point. The dimensions of the pipes and the initial conditions are listed in Table 1. The experimental results are shown in Table 2.

The relationship between water velocity and hydraulic gradient is closely associated with the internal structure of the model, as shown in Figure 4. If a karst aquifer were

Table 1. Dimensions of selected pipes and initial conditions for the confluent flow experiments.

Analog model	I	II	III
Inner diameter of pipes (cm)	1	0.6	0.8
Cross section (cm^2)	0.78	0.28	0.5
Length (cm)	85	100	160
Initial discharge (l/h)	70.4	40.4	156.1
Initial water level (cm)	31.11	22.8	37.78
Number of pipes	2	2	3

Table 2. Experimental results from the analog models.

Analog model I		Analog model II		Analog model III	
Water head (cm)	Flow rate (l/h)	Water head (cm)	Flow rate (l/h)	Water head (cm)	Flow rate (l/h)
31.11	70.4	22.90	30.4	37.78	117.6
30.42	69.2	22.21	29.6	36.45	114.7
26.83	53.6	18.62	27.1	34.72	104.4
26.14	53.1	17.93	26.6	34.14	99.0
25.36	52.6	17.42	26.1	33.39	89.6
25.06	52.1	16.85	25.6	30.37	80.4
22.28	49.7	14.07	23.2	26.27	72.6
18.82	47.1	10.61	19.8	24.88	69.7
11.86	40.8	3.65	11.2	28.33	56.3
11.02	39.5	2.81	9.5	27.32	44.9
10.25	38.4	2.04	8.0	22.17	39.8
10.20	38.3	1.99	7.9	22.00	39.5
9.74	37.3	1.53	6.4	20.74	38.0

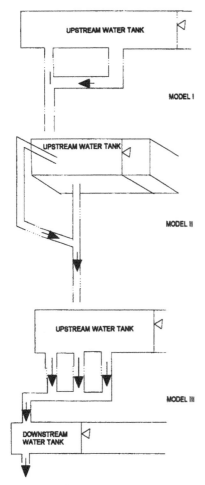

Figure 3. Analog models to study confluent flow.

made up of the three models, there would be no correlation between the two parameters. Likewise, when tracing tests are conducted within different parts of a karst aquifer with different internal structures, it is very unlikely to obtain a linear relationship between groundwater flow velocity and hydraulic gradient. Regression analyses of all the data led to:

$$v = h * 0.146 e^{-1.05I} \quad r = -0.94 \tag{5}$$

For Darcy's law to be valid, the equivalent hydraulic conductivity (hydraulic conductivity required in granular aquifers for the same velocity and hydraulic gradient) calculated by dividing the velocity by the hydraulic gradient should be a constant in each specific model. As shown in Figure 5, the equivalent hydraulic conductivity tends to decrease with the increase of the hydraulic gradient. This is probably caused by the confluent flow property in the models. More conductive media correspond to smaller hydraulic gradients such as in the master conduits, while less conductive media correspond to larger hydraulic gradients such as in the upstream joints and fractures. When karst aquifers are domi-

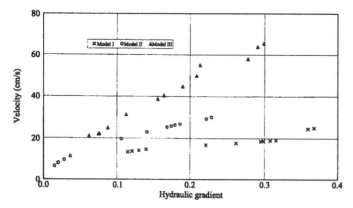

Figure 4. Water flow velocity versus hydraulic gradient from analog models

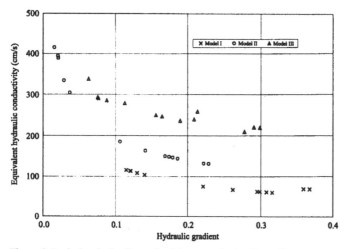

Figure 5. Equivalent hydraulic conductivity versus hydraulic gradient in analog models

nated by confluent flow, the equivalent hydraulic conductivity cannot be used to characterize the groundwater flow. When it has to be used, it cannot be treated as a constant parameter because it is related to hydraulic gradient, as indicated in Equation (5). Consequently numerical models that assume the validity of Darcy's law everywhere in the flow system should be questioned.

However, the association of the relationship between the velocity and the hydraulic gradient with the internal structure sheds a light on the solution of groundwater flow and solute transport in mature karst aquifers. In cases where the aquifers can be discretized into regions or cells where the internal structure is similar, the aquifer can be hydraulically treated separately. A network of cells can be used to represent the flow system. No assumption is made on the flow regime other than assume a confluent flow property in each cell. Mixing and flow are main factors affecting groundwater flow. Similarly, dilution and convection are main factors affecting solute transport. Diffusive flow (Darcian flow) and diffusive dispersion are neglected.

The concept of classifying karst aquifers as either 'diffuse' or 'conduit' type (Shuster & White, 1971) can also be interpreted in terms of the degree of mixing that occurs within the aquifer (Dreiss, 1989). Purely conduit-type aquifers or purely diffusive aquifers are member of the spectrum of the aquifers that actually occur in nature. In a purely diffuse flow aquifer, recharge is completely mixed with preexisting water in the aquifer, so that the spring flow shows very little variation in spring flow chemistry over time. In contrast, pure conduit flow results in little mixing and therefore large temporal variation in spring flow chemistry. The discretized cells differs from the representative elementary volume (REV) used in porous media in the following ways. (1) No constant parameters (e.g. hydraulic conductivity) can be averaged over the cell except the relationship between flow velocity and hydraulic gradient. (2) The through flow area in the cells varies significantly and can hardly be quantified. (3) The discretized cell is much bigger than the REV in porous media; and (4) No partial differential equations are readily available to describe the groundwater flow in it.

3 MIXING CELL MODELS

For several decades hydrologists have been applying mixing cell models to hydrologic systems (Dooge, 1959). Chemical engineers also have been using mixing cell models to study the efficiency of mixing in reactor vessels (Himmelblau & Bischoff, 1968). For the most part these models were solved analytically and assumed constant flow, constant cell volumes, and simple networks, all of which placed constraints on flexibility. The first application of the mixing cell model to karst aquifer was conducted by Simpson & Duckstein (1976). Llamas & others (1982) applied the mixing cell models to study the groundwater age distribution in Madrid Basin, Spain, followed by Campana and Simpson (1984) who studied the groundwater residence time in a portion of the Tucson Basin aquifer, south-central Arizona, U.S.A. Environmental isotopes and chemical species were incorporated into a mixing cell model to determine the spatial recharge distributions in the Pitsanyane and Nnywane Basins of eastern Botswana (De Vries & Gieske, 1990). Campana & Mahin (1985), using spatial and temporal tritium data obtained from samples of precipitation, stream water, and groundwater, and river recharge and spring discharge data, constructed a 34-cell model of the Edwards limestone aquifer of south-central Texas, U.S.A.. This model was non-steady state, constant volume and was calibrated by tracking the progress of the bomb-produced tritium pulse. Coupling with the three-dimensional finite-difference groundwater flow model (MODFLOW), a three-dimensional mixing cell was also developed to predict chloride concentration changes in the city of El Paso, Texas (Rao & Hathaway, 1990).

A mixing cell model computes concentration changes resulting from the mixing of waters and convective transport of solutes. The finite difference scheme for convection in the mixing cell model is described in Figure 6. For simplicity, a tracer is introduced in the model, which indicates any measurable substances dissolved in water. It can be a numerical tracer, whose magnitude is related the mass or concentration of the real world tracer. Tracers in the model and in the real world are assumed to occupy zero volume. Based on the material balance in each cell, the change in the amount of solute in a cell during a time step equals solute flux into the cell minus solute flux out of the cell:

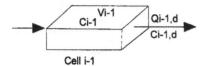

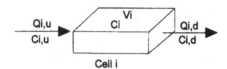

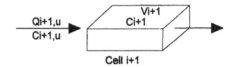

Figure 6. Schematic of three-cell one-dimensional finite-difference mixing cell model.

$$C_i^N V_i^N = C_i^{N-1} V_i^{N-1} + \left(Q_{i,u}^N C_{i,u}^N - Q_{i,d}^N C_{i,d}^N \right) \Delta t^N \tag{6}$$

where C is the tracer concentration in the cell, Q is the flow rate, superscript N is the time-step number, Δt is the increment of time, subscript i is the cell number, subscripts u and d represent upstream and downstream of a cell respectively, and V is the water volume of an individual cell, which is calculated by:

$$V_i^N = V_i^{N-1} + \left(Q_{i,u}^N - Q_{i,d}^N \right) \Delta t^N \tag{7}$$

Equations (6) and (7) are applied sequentially to each cell in the network, so that boundary discharges and concentrations from 'upstream' cells become boundary recharges and concentrations to the 'downstream' cells:

$$Q_{i,u}^N = Q_{i-1,d}^N \quad C_{i,u}^N = C_{i-1,d}^N \tag{8}$$

At the end of each time step, the water volumes and tracer concentrations in the cells are calculated. The recharge fluxes into each cell include not only fluxes through the sides of the cell but also fluxes from sources and sinks such as wells, drains, rivers, evapotranspiration, general head boundaries and aerial recharge.

In Equation (6), the only unknown on the right-hand side is $C_{i,d}^N$. This quantity can be obtained by specifying one of two mixing cell rules: the simple mixing rule or the modified mixing rule (Simpson, 1988). The former rule simulates perfect mixing within a cell; the latter simulates some regime between perfect mixing and pure piston-flow. For the simple mixing rule, the expression for $C_{i,d}^N$ is:

$$C_{i,d}^N = \left[C_i^{N-1} V_i^{N-1} + Q_{i,\mu}^N C_{i,\mu}^N \Delta t^N \right] \left[V_i^N + Q_{i,\mu}^N \Delta t^N \right] \tag{9}$$

For the modified mixing rule, the expression for is:

$$C_{i,d}^N = C_i^{N-1} V_i^{N-1} / V_i^N \tag{10}$$

The volume of water in a cell can remain constant or vary with time. Each cell represents a region of the physical system determined from the uniformity of the region of the system and the data availability. Cells may be arranged in any manner in space, so that the model may represent one-, two-, or three-dimensional system. Within each cell, either perfect mixing or some regime between pure piston-flow and perfect mixing can be simulated. Specification of pure piston-flow in individual cells does not imply pure-piston flow for the system as a whole. The concentration of the tracer in a cell is uniform, but concentration gradients may exist between adjacent cells. Mixing cell models are not hydraulic models because they do not utilize the rigorous equations of flow and do not describe the physical properties of the system. Nor are they predictive models unless they are combined with other models; rather, they are interpretive models. Water volume in each cell and the number of cells are two important parameters to be calibrated. From properly calibrated mixing cell models, information such as recharge rates, residence times, storage volumes and general flow directions can be obtained.

4 INTERPRETATION OF DYE TRACES USING MIXING CELL MODELS

A quantitative dye tracing conducted at eastern Tennessee during a storm event was used to illustrate the application of mixing cell models. About twenty-four grams of Rhodamine WT (RWT) were introduced into a sinkhole. The sinkhole drains approximately sixty acres of stormwater runoff. Stormwater and the injected RWT joined the existing groundwater in the underlying conduit and passed through two small karst windows and subsequently discharged at a spring approximately 130 m away from the sinkhole (Stephenson et al., 1997).

Dye recovery was monitored at the spring and water samples were analyzed with a fluorometer. Figure 7 shows the RWT breakthrough curve and the spring and runoff hydrograph for the period of dye trace. The breakthrough curve is the result of dye mixing with the groundwater. However, the 40 minutes lag of the peak concentration of RWT behind the peak discharge of the spring flow indicates imperfect mixing. The immediate response of spring discharge to the stormwater runoff reveals the presence of a significant segment of phreatic (completely water-filled) conduit. The hydrostatic head imposed by the stormwater inflow in the sinkhole was instantaneously propagated through the conduit and affected the spring discharge (Ashton, 1966). The turbidity that obscured the dye color at the spring indicates turbulent flow within the conduit.

Non-steady state, constant volume mixing cell models were constructed to interpret the dye trace. The average base flow of the spring was approximately 235 l/h, which was calculated by graphic interpretation of the spring discharge hydrography (Viessmann et al., 1989). The tracer concentration and the spreadness of the breakthrough curve were interactively affected by the water volume assigned to each cell and the number of cells

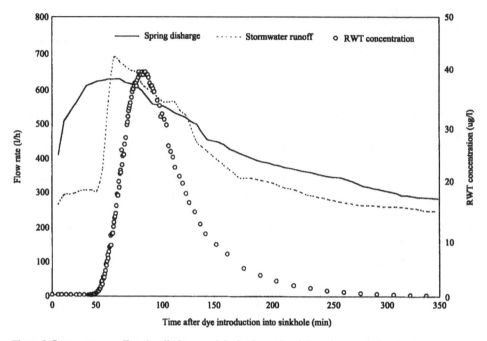

Figure 7. Stormwater runoff, spring discharge and rhodamine wt breakthrough curve during dye trace

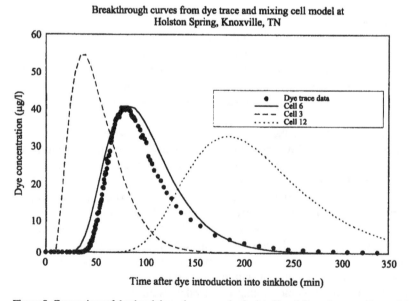

Figure 8. Comparison of dye breakthrough curve and calculated breakthrough curves from mixing cell models

in the model. The time step affected also the accuracy of the simulation and the dye breakthrough time. By assuming a complete mixing in each cell, Figure 8 compares three models consisting of 3, 6 and 12 cells, respectively. The water volume of each cell is 75 m^3 and the time step used is 5 minutes. The model with 6 mixing cells gives the best fit to the observed RWT breakthrough curve. Thus, the total conduit volume is about 450 m^3. For a linear passage, the cross-section of the conduit is about 3.5 m^2, which is not uncommon in the general area.

5 CONCLUSIONS

Comprehensive theoretical models for mature karst aquifers have not fully developed yet in spite of the substantial progress that has been made in karst research. Confluent flow regime rather than diffusive flow is probably the main obstacle. Three analog models revealed a non-linear relationship between water flow velocity and hydraulic gradient, as suggested from field tracer tests. Finite-difference mixing cell models seem to be an alternative to simulate groundwater flow and solute transport because they assume only the mass conservation. Two mixing rules have been specified to cope with different situations. The developed model was applied to interpret a dye tracer test. By adjusting the number of cells, the water volume of each cell, and the time step, a reasonable fit was achieved when the model consists of six cells with water volume of 75 m^3 in each. The design of the finite-difference mixing cell model and the initial guess of flow circulation require some a priori knowledge of geologic and hydrologic conditions. The interpreted results from the mixing cell model have not been verified.

REFERENCES

Ashton, K. 1966. The analysis of flow data from karst drainage systems. *Transaction of Cave Research Group of Great Britain*, 7: 161-203.

Campana, M.E. 1987. Generation of groundwater age distributions. *Ground Water*, 25(1): 51-58.

Campana, M.E. & Mahin, D.A. 1985. Model-derived estimates of groundwater mean ages, recharges rates, effective porosities and storage in a limestone aquifer. *Journal of Hydrology*, 76: 247-264.

Campana M.E. & Simpson, E.S. 1984. Groundwater residence times and recharge rates using a discrete-state compartment model and ^{14}C data. *Journal of Hydrology*, 72:171-185.

De Vries, J.J. & Gieske, A. 1990. A simple chloride balance routing method to regionalize groundwater recharge: a case study in semi-arid Botswana in *Proceedings, The International Symposium On Regionalization in Hydrology, Yugoslavia*, International Association of Hydrological Sciences, p.33-44.

Dooge, J.C.I. 1959. A general theory of the unit hydrograph. *Journal of Geophysics Research*, 14: 241-256.

Dreiss, S.J. 1989. Regional scale transport in a karst aquifer 2, Linear systems and time moment analysis. *Water Resources Research*, 25(1): 126-134.

Ewers, R.O. 1978. A model for the development of broadscale networks of groundwater flow in steeply dipping carbonate aquifers. *Transaction of British Cave Research Association*, 5: 121-125.

Ford, D. & Williams, W.P. 1989. *Karst geomorphology and hydrology*. Unwin Hyman Ltd, Mass. USA, 601p.

Freeze, R.A. & Cherry, J.A. 1979. *Groundwater*: Englewood Cliffs, New Jersey: Prentice-Hall, 604p.

Himmelblau, D.M. & Bischoff, K.B. 1968. *Process analysis and simulation: deterministic systems*: New York: John Wiley & Sons, 348p.

Llamas, M.R., Simpson, E.S. & Martinez Alfaro, P.E. 1982. Groundwater age distribution in Madrid Basin, Spain. *Groundwater*, 20(6): 688-695.

Milanovic, P.T. 1981. Karst hydrogeology. Colorado, *Water Resources Publications,* 434p.

Quinlan, J.F. 1990. Special problems of ground-water monitoring in karst terranes in Nielsen, D.M. & Johnson, A.I. (eds) *Ground water and vadose zone monitoring,* p. 275-304.

Rao, B.K. & Hathaway, D.L. 1990. A three-dimensional mixing cell solute transport model and its application. *Ground Water,* 28: 509-516.

Simpson, E.S. 1988. The discrete state compartment model and its applications to flow through karstic aquifers in Yuan, D.X. (ed.) *Karst hydrogeology and karst environment protection.* Guilin, China, p. 671-676.

Simpson, E.S. & Duckstein, L. 1976. Finite-state mixing cell models, in Yevjevich, V. (ed.) Karst hydrology and water resources. Ft. Collins, CO., *Water Resources Publications,* 2: 489-508.

Shuster, E. T. & White, W.B. 1971. Seasonal fluctuations in the chemistry of limestone springs: A possible means for characterizing carbonate aquifers. *Journal of Hydrology,* 14: 93-128.

Smart, P.L. & Hobbs, S.L. 1986. Characterization of carbonate aquifers. A conceptual base, in *Proceedings, The Environmental Problems in Karst Terranes and Their Solution.* Bowing Green, Kentucky, National Water Well Association, p.1-4.

Stephenson, B.J., Zhou, W.F., Beck, B.F. & Green, T.S. 1997. Highway stormwater runoff in karst areas – Preliminary results of baseline monitoring and design of a treatment system for a sinkhole in Knoxville, Tennessee, in Beck, B.F. & Stephenson, J.B. (eds) *Proceedings, The Sixth Multidisciplinary Conference on Sinkholes and the Engineering and Environmental Impacts of Karst.* Springfield: Missouri, p. 173-181.

Teutsch, G. & Sauter, M. 1991. Groundwater modeling in karst terranes: Scale effects, data acquisition and field validation, in *Proceedings, The Third Conference on Hydrogeology, Ecology, Monitoring, and Management of Groundwater in Karst Terranes.* Nashville: Tennessee, p.17-34.

Viessman, Jr.W., Lewis, G.L. & Knapp, J.W. 1989. *Introduction to hydrology* (3rd edition). New York: Harper & Row Publishers, 780p.

White, W.B. 1977. Conceptual models for limestones. *Groundwater,* 7:15-21.

Zhou, W.F. 1990. Control of groundwater from Ordovician Limestone in North China's coal fields. *Journal of China Coal Society,* 15(1): 55-66.

Zou, C.J. 1994. *Karst engineering geology of hydraulics and hydroelectricity.* Beijing, China, Hydraulics and Hydroelectricity Publishing House, 383p.

A method for producing 3-d geometric and parameter models of saturated cave systems with a discussion of applications

TODD R. KINCAID
Hazlett-Kincaid Consulting, Reading, PA, USA

ABSTRACT: A method is presented for producing interactive three-dimensional geometric and parameter models of saturated cave systems from survey data. Two and three-dimensional gridding is used to construct the outer surface of the caves and to describe the spatial distribution of a measured parameter throughout the modeled cave system. Three computer programs are described that convert survey data into Cartesian coordinates that are used to produce two and three-dimensional gridded data files. The three-dimensional models are produced from the gridded data with Dynamic Graphics' EarthVision modeling software.

The three-dimensional cave models visually convey the morphology of a conduit system in as much detail as was recorded by the survey. The EarthVision software allows multiple data sets to be incorporated into a single model from which correlations can be made between conduit morphology and various other types of hydrogeologic data such as topography, water table elevation, overburden thickness, stratigraphy, structure, etc. The parameter modeling feature will contour and display any type of discrete data measured throughout a cave system such as temperature, water velocity, pH, dissolved oxygen, and ion concentrations. Cave volumes and consequently dissolution porosities are calculated from the cave model. Examples are provided of five models developed for saturated caves in the Taurus Mountain and Antalya Travertine aquifers of southern Turkey and the Floridan aquifer of north Florida.

1 INTRODUCTION

1.1 *3-d model vs. 2-d map*

Three-dimensional models are becoming widely used in the earth sciences because they can make use of large amounts of data to display spatial relationships and variation that cannot be seen on standard two-dimensional maps (Jones & Leonard, 1990; Fried & Leonard, 1990). In karst research, two-dimensional cave maps offer limited ability to visualize and thus interpret the data they contain. Conduit depths and dimensions, even when accurately surveyed must be inferred from various labels or symbols on a standard map. Because caves are three-dimensional, understanding conduit morphology and deciphering correlations with hydrologic and geologic variables is made more difficult by the

A

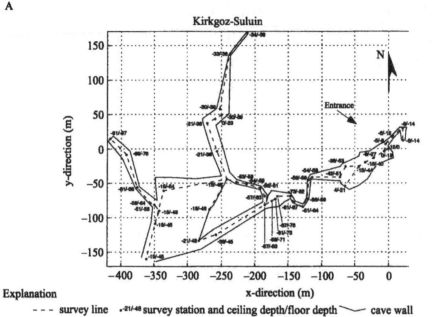

Figure 1. Map and model of the Kirkgoz-Suluin cave system developed in the Taurus Mountain aquifer of southern Turkey. (A) Two-dimensional map showing conduit width and trend. Depth of floor and ceiling are labeled at each survey station. (B) Three-dimensional model showing conduit trend and diameter where color divisions depict changes in water depth throughout the system.

simplicity of two-dimensional maps. In contrast, a three-dimensional model visually conveys the complete morphology of a conduit system in as much detail as was recorded by the survey. Figure 1 shows both a two-dimensional map and a three-dimensional model of the Kirkgoz-Suluin cave system in the Taurus Mountain aquifer of southern Turkey. Both the map and model contain the same data. Notice that the morphology of the cave system is much more obvious in the modeled depiction.

1.2 *Purpose*

This paper outlines a method for producing interactive three-dimensional geometric and parameter models of saturated caves from sparse survey data. The purpose of this paper is to demonstrate (1) the utility of sparse survey data for modeling purposes and (2) the ability of commercially available modeling software to capitalize on that data and produce viable three-dimensional cave models from which other hydrogeologic data sets can be compared and discrete parameter measurements collected from within a cave system can be visualized. A parameter model is a four-dimensional model that contours discrete measurements such as temperature, water velocity, pH, dissolved oxygen, and ion concentrations in three-dimensional space.

1.3 *Background*

Creating three-dimensional models of cave systems from survey data has been the interest of several cavers and researchers for many years (Ulfeldt, 1975; Wefer, 1983; 1990; Schaecher, 1986; Fish, 1996; Heller, 1996; Crowell, 1997; Herron, 1997; Petrie, 1997; Van Ieperen, 1997). Wefer (1989) provides a good overview and history of computerized cave mapping and three-dimensional cave modeling. Existing modeling approaches have involved the construction of custom software programs that create 3-d wire frame cave models using a series of polygons or triangles to represent conduit morphology. Most of these modeling efforts have focused on dry cave systems and capitalized on the ability of cave surveyors to collect detailed measurements of conduit cross-sections at each survey station. In this regard, the models presented in this paper are not unique except that they depict extensive saturated cave systems from which the collection of extremely detailed survey data is nearly impossible given practicable time constraints. The modeling technique described in this paper, however, capitalizes on commercially available software that utilizes gridded data sets to not only produce viable models of conduit morphology but also estimate cave volumes and render comparisons of the final model with any other two or three-dimensionally gridded data immediately attainable in a format compatible with widely used GIS systems.

1.4 *Overview*

The first part of this paper discusses the four steps involved with the modeling procedure. They are 1. the cave survey; 2. transformation of the raw survey data to Cartesian coordinates; 3. two and three-dimensional gridding; and 4. production of the final model. Procedures are described for using EarthVision modeling software (Dynamic Graphics Inc., 1998) to 1. calculate the necessary two and three-dimensional grids from the Cartesian coordinates obtained from the cave surveys and 2. produce the desired output models.

The second part of this paper presents three applications: 1. correlating conduit morphology with hydrogeologic variables; 2. contouring data trends throughout a cave system; and 3. calculating block-scale dissolution porosities. Examples are provided of five models calculated from saturated caves developed in the Taurus Mountain and Antalya Travertine aquifers of southern Turkey and the Floridan aquifer of north Florida. A map of local topography is superimposed onto a model of the Wakulla Spring cave system in the Floridan aquifer where elevation is color contoured to demonstrate correlative applications. In the remaining models, water depth is color contoured to demonstrate parameter modeling applications. Cave volumes and corresponding dissolution porosities are calculated for each of the five examples.

2 PRODUCING THE CAVE MODEL

2.1 *Cave survey*

The first step in the modeling procedure is the initial survey where the accuracy and detail of the survey define the quality of the model. In order to model conduit morphology, detailed measurements of conduit trends and dimensions must be collected. In order to model additional data collected from the cave, the location of data measurements must be recorded relative to the survey stations. The models presented in this paper are for surveys conducted in underwater caves where unforgiving time constraints limit the detail of the surveys. Greater accuracy and detail can be expected from surveys conducted in dry caves where time constraints are removed.

Decompression requirements and the diver's breathing gas supply are the two primary factors that limit the time allotted for cave penetration and survey. At the depths encountered in each of the five caves discussed in this paper, decompression requirements grow exponentially with dive time. For example, a dive to 80 m lasting 30 minutes requires approximately 110 minutes of decompression before the diver can return safely to the surface. A dive to the same depth for one hour requires approximately 260 minutes of decompression and extending the dive time out to 90 minutes increases the decompression requirement to approximately 440 minutes. Most underwater cave surveys utilize self-contained underwater breathing apparatus (scuba) cylinders to supply breathing gas for both the dives and decompressions where extended dive times require numerous scuba cylinders for both the dives and the decompressions.

Sixteen dives over the course of two summers were conducted into the four caves in Turkey. Logistical problems associated with moving scuba equipment to Turkey and then into the mountains limited the number of scuba cylinders available for the dives. Available breathing gas supplies permitted maximum dive times of 60 minutes. Multiple scuba cylinders and diver propulsion vehicles were utilized to maximize the productivity of each dive in which 30-40 minutes was allotted for surveying.

Logistical problems are greatly reduced for exploration into U.S. caves where equipment and trained cave-divers are more available. Data for the Wakulla Spring cave system in the Floridan aquifer was obtained from Stone (1989) with the results of modern exploration and surveys provided by the Woodville Karst Plain Project (Woodville Karst Plain Project, 1998). Over two hundred dives have been conducted in which 50-100 minutes were used on each dive to collect survey data. The result of the prolonged effort is greater accuracy and detail in the surveys.

Specific cave survey and mapping techniques are thoroughly documented by Hosley (1971), Thompson & Taylor (1991), and Ellis (1988). Survey techniques used for saturated caves and followed here are described by Exley (1973) and Burge (1988). The data collected consists of depth, width, and height of the conduit at discrete survey stations and the azimuth and distance between stations. Depth was recorded with a digital pressure transducer. Distance was measured with a continuous line knotted every three meters. To expedite the surveys, the orientation of the width measurement at each survey station was considered to be perpendicular to the azimuth of the forward survey direction.

2.2 *Coordinate system*

Typical cave survey data are recorded with moving polar coordinates that describe the location of each survey station relative to the one preceding it. Most computer modeling and simulation programs, however, are designed to accept data described by Cartesian coordinates. Therefore, a useful coordinate system must be selected in which to describe the spatial position of a cave system. The choice depends on the type of analyses and comparisons desired for the model. After the model is finished, coordinate transformations can be applied, if necessary, to correlate the model with other data sets.

I have chosen to position each cave in a system of Cartesian coordinates where the point (0, 0, 0) is at the entrance of the cave at the water surface. Using Cartesian coordinates for each cave system simplified the calculations. Assigning the origin to the cave entrance facilitated correlations with other map data. The water surface was used as an elevation datum because it was not possible to accurately survey cave entrance elevations and it provided a convenient datum from which to contour water depth with the parameter modeling feature. The water surface at Wakulla Spring fluctuates slightly around the zero elevation contour therefore the water surface datum was used to estimate elevation contours for that model.

2.3 *Converting the survey data into Cartesian coordinates*

The modeling procedures employed here require four separate data sets in Cartesian coordinates for each cave system that describe 1. the top part of the cave, 2. the bottom part of the cave, 3. a two-dimensional map of the outermost perimeter of the cave, and 4. the locations of each data measurement if a fourth parameter was measured. Proceeding into the cave from the entrance, the cross-sectional profile of the conduit at each cave survey station is described by four (X,Y,Z) points: top, bottom, left wall, and right wall (Fig. 2a). Together, all of the top, left wall, and right wall points describe the top part of the cave system. All of the left wall and right wall (X,Y) points describe the projection of the outermost perimeter of the cave onto a two-dimensional plane (Fig. 2b). All of the bottom, left wall, and right wall points describe the bottom part of the cave system (Fig. 2c). Data measurement locations are described by (X,Y,Z,P) points where P is the measurement value. For these models, P is equal to Z.

There are several mapping programs available that will convert survey data into Cartesian coordinates: Caps (Crowell, 1997), Cave Plot (Herron, 1997), Compass (Fish, 1996), On Station (Van Ieperen, 1997), and Win Karst (Petrie, 1997). After those programs are

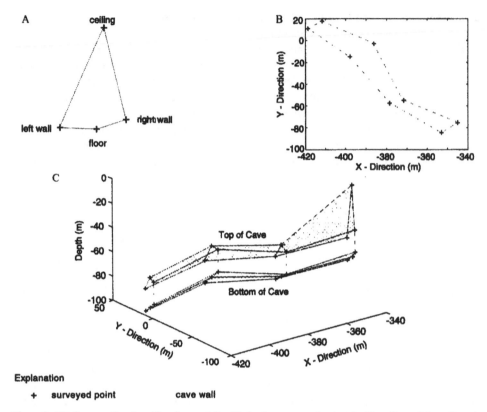

Figure 2. (A) Cross-sectional profile of a conduit with the four surveyed points (ceiling, floor, right wall, and left wall) labeled. (B) Two-dimensional projection of the outermost perimeter of a cave described by surveyed wall coordinates. (C) Multiple cross-sectional profiles describing the top and bottom of a cave separated for clarity.

used, however, the output data must be reorganized into the four required files. Data gaps between distant survey stations also have to be rectified before generating the necessary grids with the EarthVision software.

In order to produce as many points as possible for the gridding procedures and avoid complications with separating individual data sets, I designed three programs (*Cavemap.m, Expand.m,* and *Perim.m*) to convert cave survey data into the four required sets of Cartesian coordinates. Each of the programs was written for Matlab, a mathematical computer language with graphic capabilities designed by Mathworks Inc. Program codes and explicit user instructions as well as color versions of the models depicted in this paper are provided on the Internet via a persistent URL (http://research.gg.uwyo.edu/Kincaid/3dcave.htm). Program codes and instructions will also be provided by the author on request.

Cavemap.m performs the initial conversion and outputs Cartesian coordinates for the walls, floor, and ceiling of all the interconnecting conduits. *Cavemap.m* is an interactive program that requires user input to adjust or confirm conduit coordinates before producing the final output data. An initial map is produced on a local rectangular grid in either meters or feet that depicts the survey line and cross-sectional transect lines at each survey

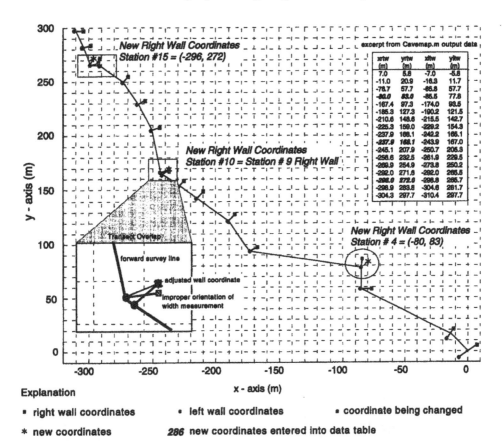

Explanation

| • right wall coordinates | • left wall coordinates | • coordinate being changed |

* new coordinates   *286* new coordinates entered into data table

Figure 3. Initial map and part of the coordinate data produced by *Cavemap.m* for the Duden Spring cave system developed in the Antalya Travertine aquifer of southern Turkey. The long line marks the cave survey while the shorter lines mark the cross-sectional transects. Overlapping wall coordinates are enclosed with boxes. The inset emphasizes a transect overlap and demonstrates one choice for adjusting the wall coordinates. An erroneous pinch-out in the 2-D profile of the conduit occurs when the assumed orientation of a width measurement closely parallels the preceeding survey line as demonstrated by the survey station enclosed with a circle.

station (Fig. 3). The graphical presentation of the input data allows the modeler to check for data errors or erroneous measurements that should be removed from the input data set before proceeding. Notice that transect overlaps occur in two locations marked with boxes on Figure 3. The output from *Cavemap.m* becomes the necessary input data for *Expand.m* that, in turn, provides the necessary data for *Perim.m*.

Expand.m expands the initial data set by linearly interpolating the positions of nine new points equidistant between consecutive survey stations (Fig. 4a). For each of the new points, *Expand.m* then interpolates ceiling, floor, and wall coordinates creating a total of thirty-six added points between each two consecutive survey stations (Fig. 4b). In effect, the data expansion is a pre-smoothing process in which a large number of points are added to the data set so that the subsequent gridding procedures will not extrapolate erroneous points between distant survey stations. The output from *Expand.m* consists of a two-dimensional map of the cave showing all of the survey and interpolated stations and

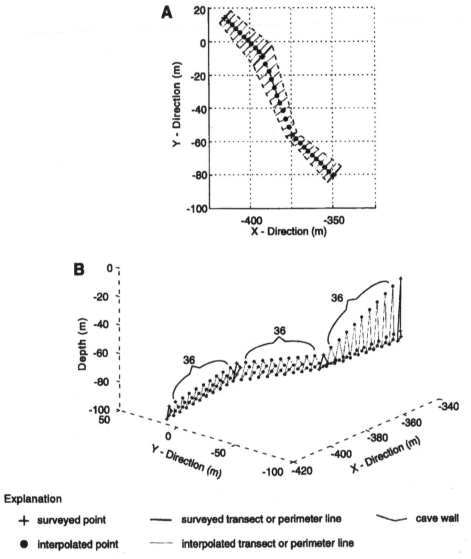

Figure 4. Graphical representation of the output from *Expand.m* showing (A) Graphical output from *Expand.m* showing the locations of all surveyed and interpolated stations and the orientation of the cross-sectional transect at each station. (B) Nine interpolated perimeter points (ceiling, floor, left wall and right wall) between the corresponding surveyed perimeter points at four survey stations. Conduit is part of the Kirkgoz-Suluin cave system developed in the Taurus Mountain aquifer of southern Turkey.

a data file consisting of the (X,Y) points that describe the two-dimensional projection of the outermost cave perimeter.

The last step in the mapping process uses *Perim.m* to approximate conduit cross-sections at each survey station. Nine additional points are interpolated along lines connecting the four cross-sectional points (ceiling, floor, left wall, and right wall) at each of

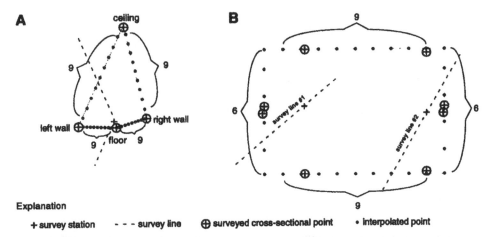

Figure 5. Graphical representation of the output from *Perim.m*. (A) Smaller conduits are surveyed along one line. Thirty-six interpolated points are created along the cross-sectional perimeter of a conduit (assumed to be an irregular polygon) at each associated survey station. (B) Very large conduits are commonly surveyed along two or more lines creating multiple sets of cross-sectional points at each associated station. Thirty interpolated points are created along the cross-sectional perimeter of the conduit (assumed to be a rectangle) at each associated survey station.

the survey and interpolated stations (Fig. 5a). The assumption underpinning the interpolations used by *Perim.m* is that the cross-sectional profile of the cave is an irregular polygon connecting the four points (ceiling, floor, left wall and right wall) at each survey station. In the case of very large conduits that were surveyed along multiple lines, the polygon is drawn as a rectangle where nine interpolated points define the ceiling and floor and six interpolated points define the right and left walls (Fig. 5b). The output from *Perim.m* consists of an initial three-dimensional plot of the data points and the three remaining data files describing the top and bottom parts of the cave and the location of the data measurements.

2.4 *Mapping problems*

Three fundamental problems obstruct the creation of a computerized map or model of a cave system from standard survey data. Loop closure errors describe erroneous gaps between subordinate conduits that connect back to a parent conduit. Loop closure errors result from inaccuracies in the survey data and can be corrected following the methods described by Wefer (1971).

Transect overlaps, depicted on Figure 3 inset, occur at closely spaced survey stations where sharp bends in a conduit cause the orientation of the width measurement to cross the plotted transect at the previous station. Erroneous pinch-outs in the cave map can also occur at single survey stations where sharp bends in the conduit cause the assumed orientation of the width measurement to nearly parallel the azimuth of the survey line as shown in the circle on Figure 3. Transect overlaps and erroneous pinchouts result from an insufficient amount of survey data to adequately describe sharp bends in a conduit. These problems can be eliminated by either collecting more survey data in problem areas or carefully measuring the cross-section of the conduit along a line that bisects the angle between straight line segments of a conduit around a sharp bend. Unfortunately, both of these methods require more time than is typically available for underwater cave surveys.

For the cave maps presented in this paper, transect overlaps and erroneous pinchouts were dealt with during the computer mapping process. The wall coordinates for overlapping transect lines were either omitted or adjusted such that a continuous line could be drawn along the wall coordinates at sequential survey stations tracing the right and left walls of the cave. As shown on Figure 3, any necessary adjustments were accomplished by reading new coordinates from the grid so that the overlap was removed. The new coordinates were then entered into the output data produced by *Cavemap.m* before expanding a data set.

2.5 *EarthVision modeling procedures*

The four files produced by *Expand.m* and *Perim.m* constitute the necessary input data for the EarthVision modeling procedures. EarthVision (Dynamic Graphics Inc., 1998) is a set of integrated gridding and visualization programs that operates on the UNIX platform. The strength of EarthVision is the capability to integrate diverse types of two and three-dimensional data into gridded graphic representations. Applications for EarthVision are currently concentrated in the petroleum industry where it is used as a tool to investigate reservoir permeability data sets. Applications are now expanding into the fields of environmental characterization and remediation, oceanographic investigations, and mining analysis. Further information about EarthVision, its capabilities, and its applications can be found in Paradis and Belcher (1990) and Belcher and Paradis (1992).

Two and three-dimensional gridding are the key concepts underpinning EarthVision's interactive modeling procedures. A spatial data set is gridded in two or three dimensions by superimposing a regular rectangular or block grid onto the data set and mathematically interpolating values at points defining the centers of the grid cells. The points defining the centers of the grid cells are called grid nodes. Grids are used as intermediate models because the mathematical rules available for interpolating or extrapolating heights on a regular grid of known values are much simpler than the rules available for use with scattered data points.

The object of a gridding procedure is to create smooth and plausible contours of spatial data in two or three dimensions by 1. honoring the input data as closely as possible and 2. calculating a plausible model for grid nodes that are not on or adjacent to input data points. Davis (1986) and Banks (1991) describe most of the commonly used interpolation techniques. EarthVision employs a minimum curvature technique that they label 'minimum tension'. This technique involves fitting a spline curve through the input data and then refining the curve to minimize its curvature that is defined by the second derivative of the spline function (Briggs, 1974).

Figure 6a shows the components of a two-dimensional grid that contains Z values (elevations, depths, or other measured data) at regularly spaced points defined by (X,Y) coordinates in a rectangular matrix. In the cave modeling procedures, two-dimensional grids contain values indicating the water depth at points along the perimeter of the cave. The depth values are interpolated by the minimum curvature gridding technique from the input data points that were generated by *Perim.m*. Two two-dimensional grids are used. One contains data describing the top part and the other contains data describing the bottom part of the cave.

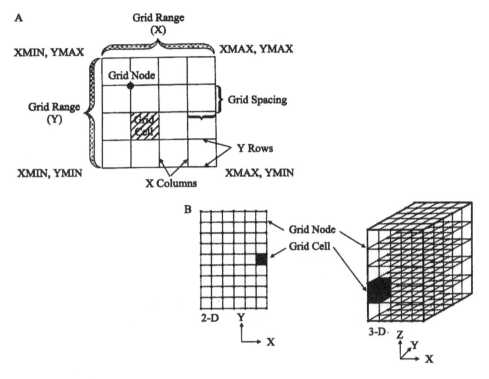

Figure 6. Components of (A) a two-dimensional grid and (B) a three-dimensional grid. Reprinted from figures provided by Dynamic Graphics Inc.

Figure 6b shows the components of a three-dimensional grid as a block model containing property (P) values that vary continuously in three-dimensional space. The P values are located at regularly spaced points defined by (X,Y,Z) coordinates in a three-dimensional matrix. The three-dimensional grid contains numbers that indicate the values of a parameter measured throughout the cave system. For the models presented here, the P values represent depth below the water surface datum. Other suitable properties that can be modeled with three-dimensional grids include temperature, water velocity, dissolved oxygen, pH, and pollutant or ion concentrations.

Five utilities in EarthVision were used to produce the cave models. They are: *3-D Minimum Tension Gridding, 2-D Minimum Tension Gridding, Faces File Generation, Graphic Editor,* and the *3-D Viewer.* Detailed instructions on the use of EarthVision software programs are not provided in this paper but are extensively covered in the EarthVision user's guide. Two other utilities, *Faces File Merging* and *Volumetrics,* were used to combine two or more modeled data sets and calculate cave volumes.

The *3-D Minimum Tension Gridding* program must be used to calculate a three-dimensional grid from the parameter data collected from the cave system. For the models presented here, the parameter data consists of depth values that were generated from the survey data by *Perim.m.* Regardless of the type of parameter data used, the data must be organized as a set of (X,Y,Z,P) coordinate points where the P-value is the parameter being contoured.

The important settings to consider are the grid spacing and the grid range. Grid spacing is expressed in the same units used to describe the cave and determines the degree of detail displayed by the output grid file. Smaller values result in greater detail but require more computation time and memory for file storage. Choice of grid spacing should be balanced against the amount of detail contained in the original survey data. The grid range is an arbitrary range defining the size of the output block. The grid range for all grids, both two and three-dimensional, used to build a single cave model must be identical.

Next, two-dimensional grids must be constructed for the top and bottom of the cave using the *2-D Minimum Tension Gridding* program. Again, the important settings are the grid spacing and grid range. After the grids have been calculated, the Graphic Editor can be used to smooth erratic surfaces that sometimes result from the gridding process. I have found that a smoothing factor of 0.5 works best to remove any anomalous irregularities without detracting from the display.

Next, the *Graphic Editor* must be used to generate a polygon file that will define the outermost perimeter of the cave. This process involves tracing the outermost perimeter of the cave from the data set produced by *Expand.m* into the polygon file with special tools provided with the *Graphic Editor*.

Finally, the *Faces File Generation* program must be used to produce the desired three-dimensional cave model from the three grids and the polygon file described above. 'Faces Files' are three-dimensional display files created from two or three-dimensional grids that can be interactively manipulated (i.e. rotated, sliced, and peeled) with EarthVision's *3-D Viewer*. The strength of the *Faces File Generation* program is the capability of clipping a three-dimensional grid with any of three different structures (upper and lower surface grids and a polygon file).

To produce a cave model, the three-dimensional grid constructed from the parameter values measured in the cave is selected as the input grid. The parameter contour interval can be set to display the desired detail. The top and bottom two-dimensional grids are used as the upper and lower clipping surfaces and the outermost perimeter data is used as the clipping polygon file. Figure 7 shows the construction of a three-dimensional model of the Wakulla Spring cave system where the clipping features are added one at a time to emphasize their function. In practice, all three clipping features are incorporated into a single run of the *Faces File Generation* program.

3 RESEARCH APPLICATIONS

3.1 *Parameter modeling*

One strength of the EarthVision modeling software is the capability to contour measured data in three dimensions and display both the data trends and conduit morphology from various angles and inclinations. Figures 8-11 depict three-dimensional models of the four cave systems in Turkey. Depth below the local water surface datum is contoured with shades of gray. Temperature, pollutant and ion concentrations, pH, dissolved oxygen, and water velocity are other types of data suited for this modeling technique. Each of the models is shown from two separate aspects of model rotation and inclination. Model rotations were accomplished in real time.

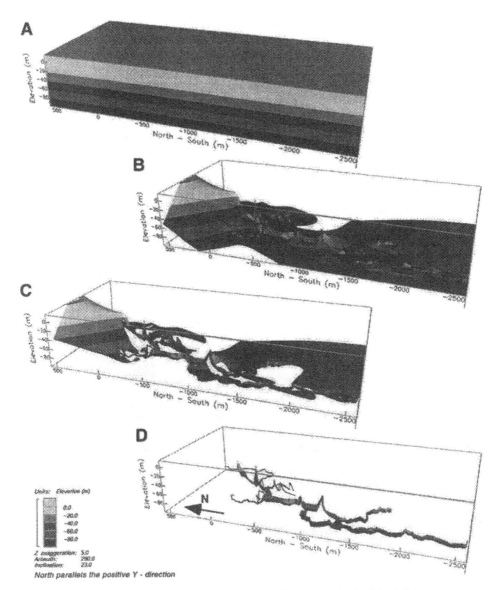

Figure 7. Calculation of a faces file with the EarthVision modeling software for the Wakulla Spring cave system developed in the Floridan aquifer, north Florida. Shades of gray reflect contours of the modeled parameter (depth). The calculation involves (A) a three-dimensional grid of depth values measured in the cave system; (B) clipping with a two-dimensional grid of points describing the upper part of the cave system; (C) clipping with a two-dimensional grid of points describing the lower part of the cave system; and (D) clipping with a polygon constructed from points describing the outermost perimeter of the cave system.

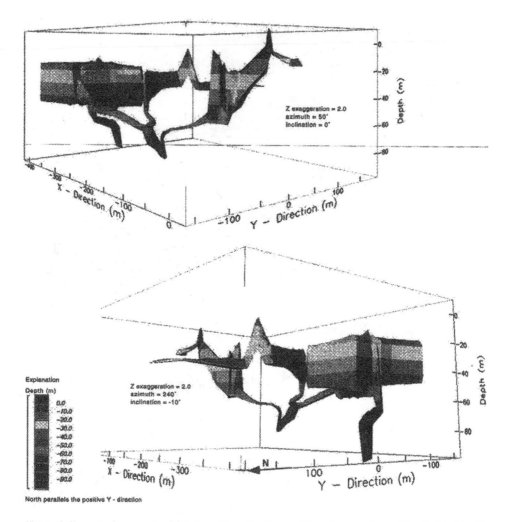

Figure 8. Two viewing aspects of the three-dimensional geometric and parameter model calculated for the Kirkgoz-Suluin cave system developed in the Taurus Mountain aquifer of southern Turkey. Shades of gray reflect changes in water depth.

3.2 *Correlating multiple data sets*

Another strength of the EarthVision software and this modeling technique is the capability to combine multiple data sets into a single model. For instance, Figure 12 shows three viewing aspects of a model of the Wakulla Spring cave system with a grid of the local topography draped over top. In this model, elevation with respect to mean sea level is contoured with shades of gray. As shown on Figure 12, a cursor can be used to correlate cave locations with overlying topographic features. The cursor shown on Figure 12 identifies a sharp rise in the conduit beneath an apparent doline marked by a depression in the topography. Regions particularly sensitive to rapid recharge can also be identified by isolating

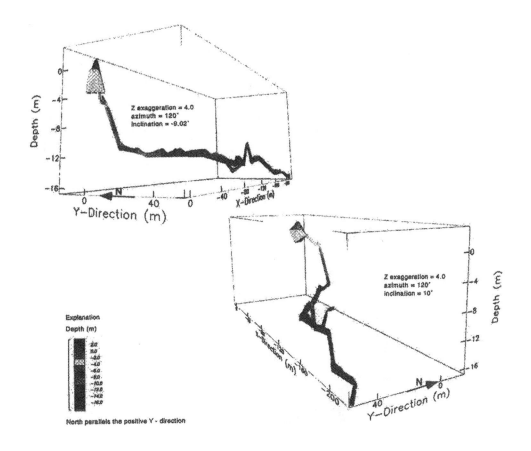

Figure 9. Two viewing aspects of the three-dimensional geometric and parameter model calculated for the Kirkgoz-1 spring cave system developed in the Taurus Mountain aquifer of southern Turkey. Shades of gray reflect changes in water depth.

3.3 *Volumetric calculations*

Cave volume calculations and consequently dissolution porosity estimates are another capability offered by this modeling technique and gridded data files. Table 1 lists the volumes of each of the five cave systems shown on Figures 8-12 and the corresponding dissolution porosity estimates.

Volumes were calculated from the cave models with EarthVision's *Volumetrics* program. The program calculates volumes by first determining average heights for the two-dimensional grids defining the top and bottom of a cave system. Then, the bottom grid is subtracted from the top grid within the area defined by the polygon constructed from the outermost perimeter coordinates. Once the cave volumes were calculated, dissolution porosity was estimated for the three-dimensional block containing a cave system using the standard equation:

$$\phi = Vc/Vb$$

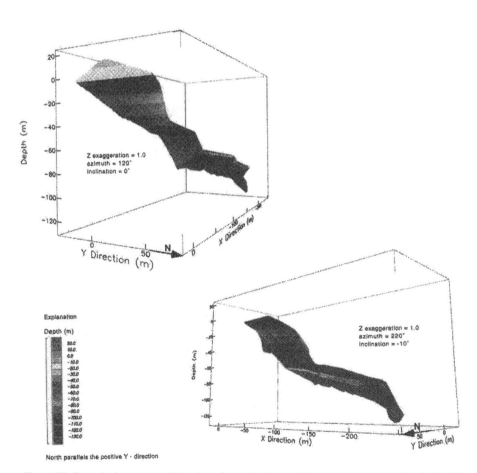

Figure 10. Two viewing aspects of the three-dimensional geometric and parameter model calculated for the Finike-Suluin cave system developed in the Taurus Mountain aquifer of southern Turkey. Shades of Gray reflect changes in water depth.

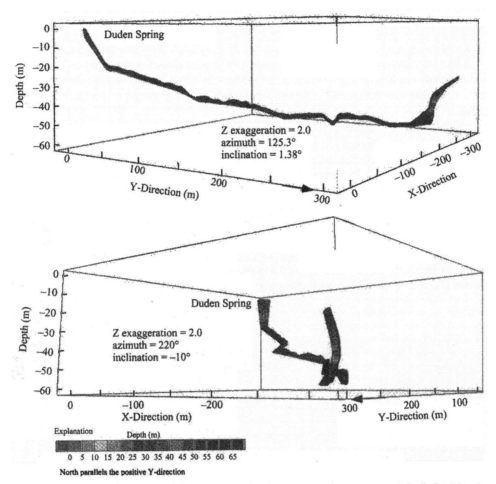

Figure 11. Two viewing aspects of the three-dimensional geometric and parameter model calculated for the Duden spring cave system developed in the Antalya Travertine aquifer of southern Turkey. Shades of gray reflect changes in water depth.

where: ϕ = dissolution porosity, Vc = volume of the cave, and Vb = volume of the block.

For each of the caves, the size of the three-dimensional block was determined from the X, Y, and Z ranges in the input data provided by *Perim.m*. Note that calculated cave volumes are only as accurate as the initial cave survey but the strength of this program is that volumes and dissolution porosities can be quickly updated as more survey data becomes available.

3.4 *On the horizon*

Unfortunately, this modeling procedure is not widely accessible because UNIX operating systems and the EarthVision software are expensive to purchase and maintain. Rapid hardware advancements producing faster and more powerful machines are, however, opening the market for developers of three-dimensional modeling software (Flynn, 1990).

Mapping programs such as Compass, On Station, and the cave mapping programs described in this paper already include features that allow survey data to be viewed in three dimensions. In addition to EarthVision, other UNIX based software packages such as IRIS Explorer (Numerical Algorithms Group, 1997) offer powerful modeling capabilities and programs such as Surfer (Golden Software Inc., 1997) and GeoGraphix (GeoGraphix Inc., 1998) are pioneering the use of sophisticated modeling programs on the PC platform.

A fundamental weakness of traditional saturated cave surveys is a dearth of data. Cave mappers in Florida (Woodville Karst Plain Project, 1998) and in Maryland (Wakulla II project, 1998) are revolutionizing the process of mapping saturated caves by developing sonar mapping systems. These sonar mappers will ride with cave explorers constantly recording cross-sectional profiles with on-board data loggers. This sonar mapping technology promises to dramatically increase the detail and availability of saturated cave survey data by collecting nearly continuous cross-sectional measurements that involve very little effort on the part of the cave-diver. The extensive data sets produced by such systems can be immediately entered into modeling programs such as EarthVision eliminating the need for the preliminary transformations and expansions. Like the models presented here, the modeled output from these future data sets will reflect all of the detail recorded by the

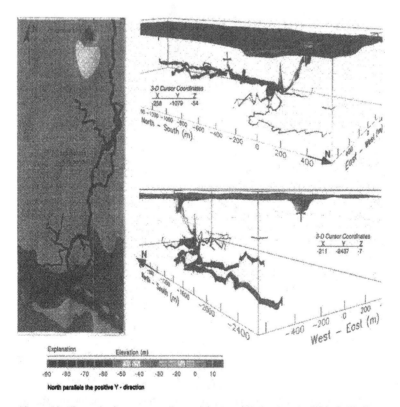

Figure 12. Three viewing aspects of a combined model showing the Wakulla Spring cave system that has developed in the Floridan aquifer of north Florida below a map of the local topography. Shades of gray reflect changes in elevation relative to mean sea level.

Table 1. Volumes and dissolution porosities calculated from models of the five cave systems shown on Figures 8-12.

Cave system	Aquifer	Cave volume (m³)	Dissolution φ
Kirkgoz-Suluin	Taurus Mountain	410,700	0.024
Kirkgoz-1 Spring	Taurus Mountain	600	0.002
Finike-Suluin	Taurus Mountain	120,000	0.036
Duden Spring	Antalya Travertine	7,100	0.001
Wakulla Spring	Floridan	650,000	0.004

better surveys and further increase the model's utility as a tool for morphologic analyses and volumetric estimations.

Another interesting prospect for the future of three-dimensional parameter modeling techniques centers on the increasing efforts to collect useful data from saturated caves. In recent years, cave diving has become more widely recognized by the scientific and regulatory communities as a useful vehicle with which to collect water samples from karst aquifers. Several groups are collecting a wide variety of biological and chemical data from deep within the saturated cave systems of the Floridan aquifer (Stoessel et al., 1989; Brigmon et al., 1994; Kincaid, 1994; Whitecross & Osmond, 1995; Banner et al., 1996). Researchers in Florida and Turkey are now using data sondes carried by a diver to continuously collect geochemical data through a cave system (Spokane, 1997; Bayari, 1996). As these types of data become more available, three-dimensional parameter models will provide the best method for visualization and interpretation.

4 DISCUSSION

My interest with these models centers on using calculated cave volumes to estimate dissolution porosity and effective permeability. Quantifying the mega-scale dissolution porosity in a karst aquifer provides a first approximation of the local effective permeability. Quantifying the effective permeability has immediate implications for regional scale ground water models used to predict ground water velocities, contaminant transport times, and ground water resource depletion.

I am finding that three-dimensionally gridded data delineating a cave system through a block of rock can be used to resolve the fractal dimension of that cave system (Kincaid, 1999). In turn, the fractal dimension describes the volume filling character of the cave (Curl, 1986). The dimension is fractional because cave systems are not perfectly three-dimensional in that they cannot be described as spheres or cubes. It appears that the fractal dimension also contains information about the conduit hydraulics and can be used to predict regional permeability in a karst aquifer from small-scale measurements.

To measure the fractal dimension of a cave, three-dimensionally gridded data are used to measure the volume of a cave relative to the coarseness with which the data is sampled. A small *sample-block* is stepped through the gridded data recording the number of times it falls completely within the cave system. The size of the sample-block defines the coarseness of inspection where resolution decreases with larger sample-blocks. Cave volume can then be measured by:

$$CV = n * \lambda$$

where CV = cave volume, n = number of sample-blocks falling within the cave system, and λ = volume of the sample-block.

A fractal dimension can then be determined for each cave system by:

$$D = \log C - \log N / \log r$$

where D = fractal dimension, C = proportionality constant equivalent to the Euclidean dimension of the grid block = 3, $norCV$ = cave volume normalized with respect to the volume of the grid block, and $nor\lambda$ = volume of the sample-block normalized to the volume of the grid block.

Fractal dimensions calculated for the five cave systems discussed in this paper range from 3.2 to 3.9. The caves with less surface discharge and diffuse sources of recharge display the highest fractal dimensions and are thus more volume filling. These data suggest that hydrogeologic factors such as local hydraulic gradients and recharge mechanisms are controlling the cave development in these regions. I am currently correlating volumetric (described above) and planar (determined from two-dimensional map data) fractal dimensions to expand these data sets and further investigate these relationships.

It is now widely recognized that permeability and hydraulic conductivity measurements scale with the volume of aquifer sampled (Bradbury & Muldoon, 1990; Smart et al., 1991; Teutsch & Sauter, 1991; Rovey & Cherkauer, 1995). Dissolved cave volumes represent an approximation of dissolution porosity encompassing the largest possible scale of a karst aquifer. Volumetric calculations from three-dimensional gridded sets of cave survey data present one method of measuring large-scale dissolution porosity. Effective permeability approximations determined from those porosities represent local maximums that cannot be measured by standard techniques.

The measurement of a maximum potential permeability is critical for accurately assessing groundwater velocities. However, it is unlikely that these values will be applicable over large regions for ground water modeling purposes because the matrix permeability is not reflected in these estimations. Moreover, it is unlikely that detailed surveys such as the ones described in this paper will always be available even if accessible, extensive conduit systems are identified. What is needed is the ability to predict regional permeability from values collected by traditional methods, keeping in mind the potential maximums.

I believe that fractal dimensions calculated for individual cave systems can be used as scaling factors to predict permeability throughout ground water catchments of karst aquifers. Assuming that the process of dissolution porosity development is similar for the largest caves and the smallest tubes, the fractal character will be similar throughout those scales of interest. Whereas traditional permeability measurement techniques yield values significantly smaller than observed aquifer characteristics, the fractal dimension of caves associated with a particular catchment can be used to scale those values to more closely describe observed conditions. When available, dissolution porosities calculated from actual cave systems by methods such as the one presented in this paper will determine the upper bound on catchment wide aquifer permeability. Otherwise, spring discharges that reflect the largest aquifer permeability for a particular catchment can be used to estimate an upper limit.

REFERENCES

Banks, R. 1991. Contouring algorithms. *Geobyte,* 6(5): 15-23.

Banner, J.L., Musgrove, M., Asmerom, Y., Edwards, R.L. & Hoff, J.A. 1996. High resolution temporal record of Holocene ground-water chemistry; Tracing links between climate and hydrology. *Geology,* 24(11): 1049-1053.

Bayari, S. 1996. International Research and Application Center for Karst Water Resources, Hacettepe University, Ankara, Turkey: Personal communication, Ankara, Turkey.

Belcher, R.C. & Paradis, A. 1992. A mapping approach to three-dimensional modeling, in Turner, A.K. (ed.) Three-dimensional modeling with geoscientific information systems: NATO ASI Series C. *Mathematical and Physical Sciences,* 354: 107-122.

Bradbury, K.R. & Muldoon, M.A. 1990. Hydraulic conductivity determinations in unlithified glacial and fluvial materials, in Nielsen, D.M. & Johnson, A.I. (eds) *Ground water and vadose zone monitoring. ASTM STP 1053: American Society for Testing and Materials,* Philadelphia, p. 138-151.

Briggs, I.C. 1974. Machine contouring using minimum curvature. *Geophysics,* 39(1): 39-48.

Brigmon, R.L., Martin, H.W., Morris, T.L., Bitton, G. & Zam, S.G. 1994. Biogeochemical ecology of *Thiothrix* spp. in underwater limestone caves. *Geomicrobiology Journal,* 12: 141-159.

Burge, J. 1988. *Basic underwater cave surveying: National Speleological Society Cave Diving Section,* Branford, Florida, 127 p.

Crowell, H. 1997. *Caps:* http://pages.prodigy.com/caps/, accessed 1998, September 22.

Curl, R.L. 1986. Fractal dimensions and geometries of caves. *Mathematical Geology,* 18(8): 765-783.

Davis, J.C. 1986. *Statistics and data analysis in geology 2nd ed..* New York: John Wiley & Sons.

Dynamic Graphics Inc., 1998, *EarthVision:* http://www.dgi.com/ev.shtml, accessed 1998, April 14.

Ellis, B. 1988. *An introduction to cave surveying.* British Cave Research Association, 40 p.

Exley 1973. *Mapping underwater caves 2nd ed..* National Association for Cave Diving, Gainesville, Florida, 22 p.

Fish, L. 1996. *Compass:* http://members.iex.net/~lfish/compass.html, accessed 1998, September 22.

Flynn, J.J. 1990. 3-D computing geosciences update hardware advances set the pace for software developers. *Geobyte,* 5(1): 33-35.

Fried, Charles C. & Leonard, J.E. 1990. Petroleum 3-D models come in many flavors. *Geobyte,* 5(1): 27-30.

GeoGraphix Inc. 1998. *GeoGraphix:* http://www.geographix.com/, accessed 1998, April 14.

Golden Software Inc. 1997. *Surfer:* http://www.Goldensoftware.com/, accessed 1998, April 14.

Heller, M. 1996. *TopoRobot:* http://www.geo.unizh.ch/~heller/toporobot, accessed 1998, September, 21.

Herron, D. 1997. *Cave Plot:* http://members.aol.com/caverdave/CPHome.html, accessed 1998, September 21.

Hosley, R.J. 1971. *Cave surveying and mapping:* Crown Press, Indianapolis, 136 p.

Jones, T.A. & Leonard, J.E. 1990. *Why 3-D modeling?* Geobyte, 5(1): 25-26.

Kincaid, T.R. 1994. Groundwater and Surface Water Interactions in the Western Santa Fe River Basin near High Springs, Florida (Master's thesis) University of Florida, 189 p.

Kincaid, T.R. 1999. Morphologic and Fractal Characterization of Saturated Karstic Caves (Ph.D. Dissertation) University of Wyoming, 174 p.

Numerical Algorithms Group 1997. *Iris Explorer:* http://www.nag.co.uk/Welcome_IEC.html, accessed 1998, April 14.

Paradis, A. & Belcher, R.C. 1990. Interactive volume modeling; a new product for 3-D mapping. *Geobyte,* 5(1): 42-44.

Petrie, G. 1997. *Win Karst:* http://www.europa.com/~gp/winkarst.html, accessed 1998, September 22.

Rovey, C.W. & Cherkauer, D.S. 1995. Scale Dependency of Hydraulic Conductivity Measurements. *Ground Water,* 33(5): 769-780.

Schaecher, G.R. 1986. 3-d cartography for the rest of us. *Compass & Tape,* 4(1): 20-23.

Smart, P.L., Edwards, A.J. & Hobbs, S.L. 1991. Heterogeneity in carbonate aquifers, effects of scale, fissuration, lithology and karstification in Quinlan, J.F. & Stanley, A. (eds) *Proceedings of the third conference on hydrology, ecology, monitoring and management of ground water in karst terranes, Nashville.* U.S. Environmental Protection Agency & National Ground Water Association.

Spokane, R.B. 1997. Research Scientist for YSI Incorporated, 1725 Brannum Lane, Yellow Springs, Ohio 44387: Personal communication, Annual Meeting of the National Speleological Society Cave Diving Section, Branford, Florida.

Stoessell, R.K., Ward, W.C. & Ford, B.H. 1989. Water chemistry and CaCO₃ dissolution in the saline part of an open-flow mixing zone, coastal Yucatan Peninsula, Mexico. *Geological Society of America Bulletin*, 101(2): 159-169.

Stone, W.C. 1989. *The Wakulla Springs project*. The U.S. Deep Cave Diving Team, National Speleological Society, Huntsville, Alabama, 210 p.

Teutsch, G. & Sauter, M. 1991. Groundwater Modeling in Karst Terranes: Scale Effects, data Acquisition and Field Validation, in Quinlan, J.F. & Stanley, A. (eds) *Proceedings of the third conference on hydrology, ecology, monitoring and management of ground water in karst terranes, Nashville:* U.S. Environmental Protection Agency & National Ground Water Association, p. 17-34.

Thompson, K.C. & Taylor, R.L. 1991. The art of cave mapping. *Missouri Speleology*, 21(1-4): 182 p.

Ulfeldt, S. 1975. Computer drawn stereo three-dimensional cave maps in *Proceedings, National Cave Management Symposium, Albequerque, New Mexico*, p. 128-129.

Van Ieperen, T. 1997. *On Station*: http://onstation.com/, accessed 1998, September 22.

Wakulla II Project, *Bill Stone*, http://www.wakulla2.com/, accessed 1998, September 23.

Wefer, F.L. 1971. The cave survey computer program: *Nittany Grotto Newsletter*, 19(1): 5-22.

Wefer, F.L., Igoe, J.W. & Gillen, P.A. 1983. An application of interactive computer graphics to the study of caves. *NSS Bulletin* 45(2).

Wefer, F.L. 1989. The computerization of the cave map. *Compass & Tape*, 7(1): 3-14.

Wefer, F.L. 1990. Content definition and control for stage-4 cave maps. *Compass & Tape*, 7(4): 3-23.

Whitecross, L.R. & Osmond, J.K. 1995. Uranium isotopic tracing of karstic conduit systems. *Geological Society of America Abstracts with Programs*, 27(6): 180-181.

Woodville Karst Plain Project, *George Irvine*, http://www.wkpp.org, accessed 1998, September 22.

Keyword index

Printed and bound by CPI Group (UK) Ltd, Croydon, CR0 4YY

23/10/2024

01777686-0005